Kelly Maria Rêgo da Silva
João Luiz M. de Sousa
Ana Florise M. Oliveira

Analysis of the profile of self-medication in neighborhoods in Caxias-MA

Kelly Maria Rêgo da Silva
João Luiz M. de Sousa
Ana Florise M. Oliveira

Analysis of the profile of self-medication in neighborhoods in Caxias-MA

Cangalheiro, Salobro and Seriema

Imprint
Any brand names and product names mentioned in this book are subject to trademark, brand or patent protection and are trademarks or registered trademarks of their respective holders. The use of brand names, product names, common names, trade names, product descriptions etc. even without a particular marking in this work is in no way to be construed to mean that such names may be regarded as unrestricted in respect of trademark and brand protection legislation and could thus be used by anyone.

Cover image: www.ingimage.com

This book is a translation from the original published under ISBN 978-613-9-62738-7.

Publisher:
Sciencia Scripts
is a trademark of
Dodo Books Indian Ocean Ltd. and OmniScriptum S.R.L publishing group

120 High Road, East Finchley, London, N2 9ED, United Kingdom
Str. Armeneasca 28/1, office 1, Chisinau MD-2012, Republic of Moldova, Europe
Printed at: see last page
ISBN: 978-620-7-74960-7

SUMMARY

ACKNOWLEDGMENTS 3

SUMMARY 5

1 INTRODUCTION 6

2 THEORETICAL BASIS 9

3 METHODOLOGY 25

4 Results and discussion 27

5 CONCLUSION AND PROPOSAL 37

6 REFERENCES 38

7 APPENDICES 48

8 ANNEXES 56

I dedicate this work especially to everyone who helped me complete this course, especially my parents, family and friends, and everyone who gave me the strength to achieve this feat.

ACKNOWLEDGMENTS

To God for being the light that illuminates every day of my life;

To my dear and beloved parents, Francinilde Maria Morais Oliveira and Raimundo Nonato Oliveira, for all your love, affection, patience and perseverance in this arduous task of raising your children and giving us the best. We may not have all the riches of this earth, but you have given us what is most valuable in this world, Thank you for the education, the principles and the reference that I will carry with me throughout my life. We have been through many adversities, including the homesickness during the two years that I lived in Buenos Aires, Argentina, studying for medicine, but in the end love has always been the preponderant and unifying link. Thank you for the words of encouragement and faith that served as a foundation for my steps during my daily trips to Teresina;

I would like to thank my brother Jônathas Morais Oliveira, who has always been a mirror of determination, overcoming and strength for me;

To my aunt Maria José, uncle James Cavalcante, Camila, Amanda Gabriella, uncle Ferdinan Morais, uncle Fernando, uncle Paulo, uncle Francinaldo de Jesus Morais, Marxo and Olga and the entire Morais e Oliveira family for all the family gatherings, for our unity, for their concern for my studies, for their faith in my achievements and for being wonderful;

I would like to thank two more important people in my life, my godmother and godfather who I love so much, Jesuslene Viana and Luiz Ferreira;

I would like to thank my grandmothers, who are no longer alive, but have been present throughout this journey, Rosa Soares and Inês Oliveira, my love for you is eternal;

To my friends and colleagues, whom I have made throughout my life and also during this degree, for their encouragement and support.

To my second mother, Francilene Maria Morais, for playing an unparalleled role in the development of this work, as well as for being that precious human being full of good energy who captivates everyone at all times. Thank you for being there;

To Professor Solange Santana Guimaraes Morais, who guided me for hours and hours, but always with great generosity and patience. And to Professor Joao Luiz, my advisor, for all his dedication and commitment to my research and learning. Thank you very much;

I would like to thank all the staff at the Mauricio de Nassau College for the excellent collective work that has taken place at this college and who, directly and indirectly, have contributed to building a wonderful study and work environment;

To the residents of the Cangalheiro, Salobro and Seriema neighborhoods for participating in this research and for making it possible to learn more about each of you;

And the college will have many memories.

"Instead of taking the floor, I would like to be enveloped by it and taken far beyond any possible beginning."

(Michel Foucault).

SUMMARY

Self-medication is a common practice in Brazil and around the world, which is a cause for concern as it is a public health problem. To this end, this study aims to analyze the profile of self-medication among residents of the Cangalheiro, Salobro and Seriema neighborhoods in the city of Caxias do Maranhâo in 2016. It can be seen that there are causes that lead to self-medication, including the financial difficulties faced by a large proportion of people, trying to solve illnesses based on the opinions of other people or other sources of information, deficiencies in the public health system, excessive advertising by the pharmaceutical market through publicity campaigns where there is a lack of information about the side and adverse effects caused by the use of drugs. This work is based on the authors Loyola Filho, Damasceno, Chacra and Cassiani, who presented pertinent observations on the subject, as well as a bibliographical survey of articles from SciELO. The nature of the study was descriptive and cross-sectional observational, collecting information on self-medication and sociodemographic data. Nine hundred randomly selected people were interviewed. The Excel 2010 program and SPPS 20.0 software were used. It was concluded that there is a relationship of dependence between neighborhood and self-medication. With the results obtained, we also intend to propose measures that can minimize the consequences of this practice and mobilize awareness among the greatest number of people.

Key words: Self-medication. Public Health. Residents. Medicines. Advertising. Risks.

1   INTRODUCTION

Self-medication is a common practice in Brazil and around the world. According to Loyola Filho et al. (2005), this activity can be practiced by purchasing medication without a prescription, sharing medication with other members of the family or social circle and using leftover prescriptions, reusing old prescriptions and failing to comply with professional prescriptions, prolonging or prematurely interrupting the dosage and the period of time indicated in the prescription.

In the context of human history, self-medication has been used since the beginning of human existence. During many phases of the evolution of history, we have always sought to develop various therapeutic forms to alleviate the symptoms presented during the onset of an illness, either through natural herbicides with curative purposes, homemade medicines, galenic mixtures, and also industrialized medicines. (CRUZ; CARAMONA; GUERREIRO, 2015).

Pharmaceuticals have a close relationship with health, from delaying the development of an illness to curing it. However, when used irrationally, they can pose a risk to an individual's state of health, cause intoxication or lead to death (LIMA, NUNES; BARROS 2010).

In today's therapeutic treatment, drugs are seen as an important element, with a curative purpose and to control pathologies, with great cost-effectiveness through rational use (LEITE, VIEIRA; VEBER, 2008).

Some of the conditions that favor the expansion of this practice in the world are political, cultural and economic, and it is seen as a serious public health problem. The huge increase in supply on the market has led to a familiarity of the lay consumer with medicines (LOYOLA FILHO et al., 2005). Allied to these conditions are other factors, such as the existence of gaps in health care and the difficulty of health services in underdeveloped countries (CARVALHO; PASCOM; SOUZA-JUNIOR; DAMACENA; SZWARCWALD, 2003).

Medication errors are also defined as:

Any foreseeable incident that could cause harm to the patient or lead to inappropriate use of medicines when they are under the control of healthcare professionals or the consumer patient. These incidents can be related to professional practice, procedures or systems, including failures in prescribing, communication, labeling, packaging, naming, preparation, dispensing, distribution, administration, education, follow-up and use (LOURO et al., 2007, p. 1043).

The WHO has reported that in Brazil pharmaceuticals are marketed in large quantities, approximately 30,000 drugs a year. One possible cause of these figures is the fact that pharmacies are seen as commercial outlets rather than health facilities (SOUSA; SILVA; NETO, 2008).

According to Damasceno et al. (2007), the practice of self-medication is not without risks, as it causes side effects, often more serious than the original illness itself, and can mask serious illnesses, causing delays in diagnosis and proper medical treatment, causing serious consequences for exposed patients, as well as increasing health costs.

Marketing is the most widely used way to market products, thus promoting the purchase of items to the target audience. Based on this, the medicines trade grows exponentially every year, as the number of consumers of medicines is very large. The aim of advertising is to sell the medicine, which can detract from the real concept of medicine by advertising it as a consumer product, creating demand that exceeds the real needs of the consumer, inducing indiscriminate consumption and encouraging self-medication (CHACRA; FAGUNDES; PINTO, 2008).

The benefits of medicines that advertising campaigns publicize do not cancel out the risks that may arise from the use of these products (BUCARETCHI, 2007).

The package leaflet or guidance of a medicine used by someone else does not override the assessment of a health professional. Since pharmacotherapy requires complete treatment, regardless of whether the pathology is mild or severe (CENÇO, 2010).

The financial investment of the pharmaceutical market has shown results in recent years that foreshadow a threat to the sustainability of public health agencies in many countries. According to Mota et al. (2008), increasing spending on the supply of medicines has been competing for space with other activities in the health sector and these financial costs are not showing good results when the quality indices of this segment are evaluated.

Given this, and also the need to study cases of toxicity due to the misuse of medicines and their complexities, this study will address this issue in three traditional neighborhoods in Caxi: Cangalheiro, Salobro and Seriema.

The predefined criteria for choosing the neighborhoods and applying the survey were: the cultural and socio-economic differences, together with the variety of academic and age levels in these communities.

Some factors have contributed to the communities in these neighborhoods self-medicating, either because of financial difficulties, because they can't get places in public hospitals, or because they can't get medical or dental care,   or because they trust the advice of other people who are not qualified.

The research was based on theoretical criteria in the exploratory mode through data collected from articles published by SciELO, drug advertisements and research sites. A mixed questionnaire of twenty-six open and closed questions was applied. Among the theoreticians of this research are Antônio Ignâcio de Loyola Filho (2002), Dênis Derly Damasceno (2007) and others who contributed to the theoretical foundation of this study.

This is a simple descriptive, qualitative and quantitative study, as well as a cross-sectional observational study.

The objectives set for this study were: To analyze the risks of self-medication through a survey of qualitative and quantitative data from residents in the Cangalheiro, Salobro and Seriema neighborhoods in the city of Caxias-MA. The specific objectives were To present the main reasons why the population of the Cangalheiro, Salobro and Seriema neighborhoods in the city of Caxias-MA resort to self-medication; To show through graphs the number of people in the Cangalheiro, Salobro and Seriema neighborhoods who have knowledge about the risks of the medication they have consumed, as well as those who do not have information on the subject; Identify the consequences of self-medication and present the risks it produces; Identify the number of people who resort to self-medication in the research population.

This monograph is structured in five chapters. The first chapter deals with the scientific area of study. The second deals with the theoretical foundation and clarifies the contextualization. The third chapter highlights the methodology used to carry out the work. The fourth chapter presents the interpretation and analysis of the data collected. The fifth chapter presents the conclusion of the research and proposes suggestions for intervention to help bring about change.

The aim of this work is to contribute to an energetic reflection on the risks caused by self-medication and its consequences, as well as to help improve the public health system in Caxias-MA through the data needed to minimize the number of people who use self-medication. In addition, the participation of all citizens, whether they are political representatives or not, in collaborating with preventive and social measures through the correct use of the information obtained is essential. The conclusion of the research serves as a further source of study for all health researchers.

2 THEORETICAL BASIS

2.1 Histories of pharmacy and medicines

Since the emergence of disease, man has always sought ways to find a cure.

Ancient texts report the use of plants and substances of animal origin for curative purposes as far back as the Paleolithic period or chipped stone age. The oldest known pharmaceutical document is a Sumerian tablet (clay table) made in the third millennium (2100 BC), containing fifteen medicinal recipes, discovered in Nippur. The most important papyrus in the history of pharmacy is the Ebers papyrus written around 1500 BC, a kind of manual for students, which reveals the secrets of medicine. This veritable pharmacopoeia contains a wealth of information, 811 prescriptions and 700 remedies for various diseases, from snakebite to puerperal fever, covering[1] .

Gomes Jùnior (1988, apud PEREIRA; NASCIMENTO, 2011) states that modern medicine began in Ancient Greece with the use of apothecaries, also known as apothéke, a word of Greek origin meaning a small box for storing pharmaceuticals and used on trips made by doctors during health care.

Until the 11th century, medicine and pharmacy were a single profession, but according to Pereira and Nascimento (2011), in 1240 A.D., the Roman emperor Frederick II published the Magna Carta in which he declared pharmacy to be an independent profession.

During the colonial period, the apothecary appeared in Brazil. Allied to this, apothecaries sold medicines and other products with a curative purpose, while the apothecary usually produced and administered the drug in front of the patient according to the doctor's prescription and the pharmacopoeia. (CONSELHO REGIONAL DE FARMACIA DO ESTADO DE SAO PAULO, 2016).

According to Bittencourt et al (2003), at the end of the 19th century Pharmacology underwent its greatest development, together with the impetus of modern technologies. In the field of Physiology, in the middle of the 19th century, Claude Bernard's research was highlighted through his study of the action of curare, as well as its effect on animal muscles (PORTER, 2004, 1996; BLACK, 1999).

A milestone in Pharmacology was presented by the researcher Claude Bernard during the 20th century, which contained the criteria for collecting informative data for experimental medicine. However, the implementation of these criteria in the field of therapy is

---

1 - Available at: <http://www.sbfc.org.br/site/paginas.php?id=2>. Accessed on: September 15, 2016.

inconsistent and time-consuming (NIES, 2001, p.45).

Another important researcher in the field of pharmacology was George Urdang. In 1927, George Urdang, a pioneer in the study of the history of pharmacy, considered the pharmacist and his function as the object of a scientific discipline (DIAS, 2016).

Some events have marked the evolution of pharmaceutical science in the field of research. According to Angell (2007), during the 1980s many important changes took place in the field of pharmacology.

Studies such as those by Bermudez (1992) and Rozenfeld (1989) suggest that, in Brazil, the rational use of medicines is becoming a distant reality, both in the private and public sectors, whether in hospitals, outpatient clinics or in the community.

Pharmacoepidemiology, or drug epidemiology, emerged from the intersection of clinical pharmacology and epidemiology. The end of the 19th century and the beginning of the current century were characterized by a historic turnaround in the field of therapeutics (CASTRO, 2000, p. 15-16).

2.2 Pharmacological concepts

The process of pharmacotherapy is a way of treating human pathologies through the use of drugs. According to Anvisa (2016), medicines are pharmaceutical products produced with the aim of curing, or as a prophylactic, diagnostic or palliative measure.

Pharmaceuticals are described as complex, symbolic and representative of Western society, as they are not just a mechanism for therapeutic intercession (ALBALADEJO; DiEZ, 2002).

In the health sector, medication is important both in the system's management policies and in the practice of the professionals involved, as well as in terms of patients' emotional references (SANTOS; NITRINI, 2004, p. 820).

Another point to note is that medicines are not just any products, they act to relieve pain and also to cure pathologies. However, care must be taken when administering them, as they can be dangerous when not used safely. It is also a unique product, since it involves different stages in the production and marketing process. In order for this product to reach the consumer, it is necessary to carry out research into biologically active molecules, develop the formulation process with a view to production on an industrial scale and achieve final processing. Then the process of marketing and distribution in the commercial network and health services begins (OLIVEIRA; LABRA; BERMUDEZ, 2006).

Administering medicines to patients in healthcare settings requires a complex, multi-stage process involving a series of interrelated decisions and actions involving professionals from various fields and a team who interact one-on-one with the patient. "The professionals involved are: the doctor, pharmacist, pharmacist's assistant, nurse and nursing assistant or technician" (CASSIANI, 2005, p. 2).

The advances in research into new drugs, together with their commercial promotion, have created an excessive belief in society in relation to the power of medicines. In this way, their indication becomes almost necessary in medical consultations, and the doctor is consulted by the patient through the number of pharmaceutical forms he prescribes. Thus, prescribing medication has become synonymous with good medical practice, justifying its huge demand (SILVEIRO; GONÇALVES, 2010).

Despite all this information, there is a fundamental principle in the administration of medication, called the principle of the five right things, i.e. that the right medicine is given to the right patient, in the right dose, by the right route and at the right time (CASSIANI, 2005, p. 2).

The search for quick solutions to relieve pain or other pathological conditions is justified by the use of drugs. However, the false illusion of immediate treatment is not always the right choice, as drugs can have significant side effects and lose their effect with prolonged use. (AQUINO et al., 2010; GALATO, 2012; LOYOLA, 2002; MUSIAL, 2013; NARLOCH, 2004; PENNA, 2004; SCHUELTER et al., 2011; SOUZA, 2013).

So the idea that errors can occur is notorious at all stages of the medication system: 39% of errors occur during prescription, 12% during transcription, 11% during dispensing and 38% during administration. Nurses and pharmacists intercept 86% of medication errors related to prescribing, transcribing and dispensing errors, while only 2% are intercepted by patients. So there is no safety net for nurses when drugs are administered to patients (LEAPE, 1995).

It is also important to consider the concept of pharmaceutical alternatives in pharmacological studies. According to ANVISA (2016), pharmaceutical alternatives are medicines that have the same active ingredient, not necessarily in the same dosage, pharmaceutical form, chemical nature (ether, salt, base), but offer the same therapeutic activity.

According to Figueiras et al. (2007 apud FERNANDES et al., 2009, p. 2), he shares the view that:

To develop it, decades and millions of dollars are spent - which is why reference medicines are protected by laws that guarantee exclusivity in pharmacies for up to 20 years. On the packaging, the medicine bears an invented name - the "fantasy" name - the name of the active ingredient and the commercial name of the company that created the formula.

Generic drugs are given this fancy name because they have a higher concentration, pharmaceutical formula and indication than the reference drug. It is characterized by having a more accessible monetary value that corresponds to the target audience, on average 40% lower than the reference drug (FIGUEIRAS, 2007). The bioavailability of a reference medicine is defined during the development of the drug. It must be an innovative product with safety and efficiency proven through clinical studies before it is marketed (BRASIL, 2016).

Pain is one of the signs that something isn't right in the body. Pain plays a kind of language in which nerve cells specialized in the sense of pain (nociceptors), which exist in thousands in the skin, transmit harmful impulses to the central nervous system, which responds by trying to move the affected part of the body away from the painful stimulus (GUYTON, 2002).

Analgesics or anesthetics are drugs used to suppress pain. They can be taken orally, intravenously, subcutaneously, transdermally or intramuscularly. It is possible to distinguish three types of analgesics whose effects increase as the undesirable effects increase. Morphine derivatives are the strongest and have a rapid and/or prolonged action. Other drugs have analgesic qualities and effects in addition to their main use: antiepileptics, antidepressants or neuroepileptics. Analgesics can have side effects such as nausea, constipation or respiratory stress. Their effectiveness depends on the origin of the pain[2] .

Among them, dipyrone and paracetamol are the most widely used analgesics (AQUINO et al., 2010; SCHUELTER et al., 2011; SOUZA, 2013).

The high consumption of analgesics for headaches, flu and fever, among other symptoms, can lead to adverse effects such as gastric discomfort, ulcers and gastritis (GELLER et al., 2012; UFGRS, 2002).

Anti-inflammatory drugs are medicines that reduce the signs of inflammation such as pain, redness and heat. (DRAUZIO VARELLA, 2016)

Another concept of anti-inflammatory drugs is presented by Almeida and Silva (2013), who mention that these drugs reduce the cardinal signs of inflammation by inhibiting the production of prostaglandins, which are synthesized by the enzymes cyclooxygenase (COX) after an inflammatory stimulus in the tissues.

---

2 - Available at: < http://saude.ccm.net/faq/574-analgesicos-definicao>. Accessed on: September 15, 2016.

The use of non-steroidal anti-inflammatory drugs (NSAIDs) has become one of the most sought after drugs in drugstores due to their therapeutic actions such as analgesic, antipyretic and anti-inflammatory activities. Any kind of inflammation, muscle pain or stiff neck is a reason for people to go to a drugstore for an anti-inflammatory (LIMA; FILHO, 2010).

The process of inflammation is understood as a response by the body to cell damage and involves a large number of cells and chemical and biological mediators that trigger a complex cascade of biochemical and cellular events. This process attracts cells and stimulates the release of various intermediates, the inflammatory phase, which can include histamine, bradykinin, serotonin, arachidonic acid products and ATP (ALMEIDA; SILVA, 2013).

Another class of drugs that are very popular in pharmacies are antibiotics, which are drugs of industrial or natural origin, prepared to inhibit the growth of bacteria and fungi. They can be bactericidal, when they kill bacteria, or bacteriostatic, when they inhibit microbial growth (WALSH, 2003).

Because it is a drug that is in high demand, administration errors are very dangerous. The problem with the use of antibiotics is the development of microorganisms that are potentially resistant to any treatment, causing serious consequences for the patient and possibly leading to death (SANTOS; NITRINI, 2004).

The treatment of hypertension is achieved by reducing the cardiovascular mortality of hypertensive patients, whether fatal or not. It also reduces blood pressure levels.[3]

Hypertension has another consequence these days, as it is related to the number of elderly people in the country. As such, it is one of the current health problems, especially among the elderly (ZAITUNE et al. 2006).

According to Porto (2005, p. 482), hypertension is a syndrome that is basically characterized by an increase in blood pressure levels, both systemic and diastolic.

Arterial hypertension is one of the most important diseases in the modern world, because not only is it very common - 10 to 20% of the adult population suffer from arterial hypertension - but it is also the direct and indirect cause of a high number of deaths resulting from strokes, heart failure, kidney failure and heart attacks (PORTO, 2005, p. 487).

Syrups are viscous medicines made up of sugars, into which the active ingredient is added, causing a beneficial effect. There are cough syrups where the active ingredient is often ziperol or codeine (UNIFESP, 2016).

Coughs can be caused by bacterial or viral infections. Syrup, when used inappropriately, causes hiatus reflux and hernia reflux and can also trigger cancer in the respiratory system (IVFRJ, 2016).

Helminths cause mild or severe intestinal problems or can lead to death. Because of this, antiparasitic drugs are one of the main forms of treatment, justified by their efficiency, practicality, ease of obtaining and cost-effectiveness (MOLENTO, 2005).

De Luca et al (1999, apud FREITAS, 2006), share the concept of[3] anti-acids, which are substances that neutralize the acid projected by the stomach's parietal cells. They are medicines used in cases of hyperacidity and to treat gastroesophageal reflux and peptic ulcers. In addition, using this drug incorrectly can delay the diagnosis of tumors, pancreatitis, heart attacks and ulcers (IVFRJ, 2016).

There are also other therapeutic alternatives, such as herbal and homeopathic medicines.

Humanity has used plants as medicinal treatments and obtained beneficial results. One of the reasons given is that the use of these drugs previously did not need to undergo clinical evaluation, unlike industrial medicines (FERREIRA and PINTO, 2010). In line with this concept, Maciel et al (2002) say that medicinal plants are considered to be only those that cure or alleviate pathologies and are used frequently within a community. It is necessary to have knowledge about the plant in order to collect and prepare this medicine, and this usually happens with the help of the knowledge of the plant's traders or raizeiros.

Maclennan et al. (1996 apud, TEIXEIRA; SANTOS, 2016), believe that the main people who use herbal medicines are the elderly and adults, mainly to treat chronic diseases.

According to Anvisa (2016), homeopathic medicines are produced in accordance with the fundamentals of homeopathy, using methods of control and preparation that comply with the Brazilian Homeopathic Pharmacopoeia.

2.4 On self-medication

During the 70s and 80s, there was an expansion in the ease of access to drugs[4] , since it is considered a component of health care that is exempt from diagnosis, prevention or

---

3 - Available at: <http://departamentos.cardiol.br/dha/consenso3/capitulo5.asp>. Accessed on: August 15, 2016.
4 - World Self-Medication Industry. The story of self-care and self-medication - 40 years of progress, 1970-2010. 2010; p. 32.

treatment without consulting a doctor or dentist.[5]

Medicines are essential to health and an important therapeutic tool, accounting for a significant part of the improvement in the quality and life expectancy of the population (ARRAIS et al., 2005).

According to Singer (1993 apud CAMPOS; NETO, 2008), quality of life linked to health is the degree of importance that life should be classified and the functional complexities that can appear at any given time, such as illnesses, social conditions, perceptions, aggravated by political and economic issues of the health system and treatment.

The influx of new drugs in recent years has led to dramatic improvements in drug therapy, but it has also created problems. Not the least of these is the so-called 'therapeutic jungle', an expression used to refer to the combination of the overwhelming number of drugs, confusion over drug nomenclature and the situation of uncertainty associated with many of them... The doctor can help remedy this situation by prescribing products by their non-commercial names. More importantly, he must develop a 'way of thinking about drugs' based on pharmacological principles (GOODMAN, GILMAN, 1965, p. 33).

According to Gandolfi and Andrade (2006), medicines are necessary and relevant products for humans, along with housing, nutrition and other factors that describe the health index. However, they become controversial due to their irrational use because their therapeutic function is associated with social and economic issues, not those related to illness and health.

Self-medication is characterized as any initiative by a patient or their guardian to purchase medicines without a prescription from a legally qualified professional (LOYOLA FILHO; UCHOA, 2002). Another concept of self-medication is raised by Wertheimer; Serradell (2008), who say that this practice is the act of consuming medication without a prescription. This occurs through the patient's decision as to which medicine to use in order to cure and minimize signs through health promotion.

Since self-medication is widely practiced by Brazilians, both because of the difficulty in accessing health services and because the less privileged classes seek quick solutions to pathologies in order to prevent their daily activities from being impeded (NASCIMENTO, 2003).

Another important fact is that the number of pharmacies in Brazil is quite large and, consequently, the Brazilian population also consumes drugs in large quantities. According

---

5 - Knapp-Duglosz C. OTC Advisor - The Pharmacist's Role in Self-Care. 2009; p. 17.

15

to Cantarino (2007), Brazil is among the countries that consume the most pharmaceuticals in the world. It ranks 10th in the global pharmaceutical market. An average of 1.6 billion boxes of medicines are sold each year.

The number of people practicing self-medication is quite worrying in Brazil, according to the Brazilian Association of Pharmaceutical Industries, which claims that 80 million people have the habit of self-medicating in Brazil. Some of the reasons for this are the poor quality of medicines on offer, the fact that many pharmacies do not require a doctor's prescription when selling certain medicines and the fact that pharmacies in Brazil are not described as health facilities, but as a market for medicines (VITOR et al., 2008; SOUSA; SILVA; NETO, 2008).

In developed and developing countries, the abuse of antibiotics for the treatment of viral infections is quite common. This is due to a number of factors, such as the difficulty in discriminating between bacterial and viral infections, the preventive use of these drugs, which can cause bacterial resistance, the ineffective control of the marketing and sale of antibiotics, and the lack of information about the consequences of the improper use of antibiotics (BRICKS, 2003).

In 2003 in Brazil, the main toxic agents in humans were medicines (28.2%), venomous animals (24.1%) and sanitary domes (8.2%) (FUNDAÇÂO OSWALDO CRUZ, 2016).

As such, drug poisoning is the number one cause of poisoning in Brazil, accounting for around 18% of deaths from poisoning caused by medicines. Depressants, antidepressants and anti-inflammatory drugs are the most commonly related to cases of poisoning (ALONZO; CORRÊA, 2001).

According to SINITOX (2016), approximately 113,000 cases of human poisoning have occurred since 2006, of which medicines accounted for 30.7%.

Headaches, dyspepsia and respiratory infections are symptoms of self-medication (MUSIAL et al. 2007). Allergies are another consequence of the chemical principle of a particular drug, and can also lead to intoxication. Cerebral hemorrhage is another undesirable repercussion of this practice, caused by combining analgesics with anticoagulants (LIMA; RODRIGUES, 2016).

One of the effects of the health crisis in Brazil is self-medication. This practice is also associated with the risks of drug interactions due to polymedication, which can reduce therapeutic results and lead to serious health problems (MONTEIRO, 2016).

When patients obtain drugs according to their clinical urgency, within the reference dosage, at an affordable price and in the appropriate therapeutic timeframe, this results in the rational use of medicines. It should be emphasized that rationality is also needed with herbal and traditional medicines (PAHO, 2016).

According to Loyola Filho et al. (2005), the cultural, political and economic environment contribute to the expansion of this practice. In recent years, monetary investment has been made in the pharmaceutical market, which has led to uncertainty about the sustainability of public health in various countries. According to Mota et al. (2008) there is a dispute between the division of spending between the health sector and other sectors, which can lead to an imbalance in the quality of development indices.

"Even though it is considered a public health problem, self-medication continues to exist in Brazil. This fact can be explained by the fact that the country's health system is not prepared to meet the demand for the sale of drugs without a prescription, which is why the best option for many people is to go to the pharmacy," says Mirtes Peinado, a biomedical doctor and Idec's testing and research manager (REVISTA IDEC, 2010).

The excessive number of drug advertisements is associated with the indiscriminate use of medicines, together with the easy accessibility of these products in supermarkets and pharmacies. However, this decision exposes consumers to unwanted problems (NASCIMENTO 2003).

The highest incidence of poisoning in Brazil is caused by the improper use of medication. ANVISA records that every twenty seconds at least one person suffers drug poisoning. In order to bring about drastic changes, it is important to comply with legislation on the use of medicines, obeying the limits on the sale and release of this type of product.

(BORTOLON et al, 2008).

According to Azevedo (2010), toxicology is the science that studies the adverse consequences of chemical substances on the body and also describes the chance and probability of occurrence. Toxic agents generally related to poisoning are anticholinesterases, salicylates, metaboglobin, carboxyhemoglobin, barbiturates, cocaine and paracetamol (VIEIRA 2012).

Some of the situations related to self-medication are: using the wrong dosage, using medication for a non-corresponding pathology, disregarding the information on the package leaflet regarding contraindications, abusive dependence, drug interactions that lead to a worsening of the clinical condition. It is also worth pointing out that the elderly are the group most at risk, since they are the people who take the most polymedication, due to their need and physiological fragility, which is why cases of drug interaction in the elderly are frequent (RUIZ, 2010).

The level of schooling and information is linked to this practice, as is the ease of access to medicines in the health system (TOMASI et al, 2007).

Brazil is an emerging country and is far from being classified as a developed country, since the reality is different, especially when compared to the consumption of pharmaceuticals in first world countries. This is why there is an alarming rate of people who self-medicate in Brazil, coupled with investment from the pharmaceutical industry, which spares no effort in promoting its products through marketing and drug advertisements. Pharmacies are described as veritable supermarkets, producing a culture of unbridled consumption (JESUS, 2007).

Incorrect self-diagnosis during the practice of self-medication can lead to a more serious illness, or its concealment, which hinders the effectiveness of therapeutic treatment (HUGHES et al., 2001; RUIZ, 2010).

According to Oliveira and Munaretto (2010), the proposal of education with safe sources of information about this practice for consumers, dispensers and prescribers are measures that can minimize the appearance of resistant bacterial strains as well as preserving the efficiency of antibiotics.

1.4 Marketing and drug advertisements

Spreading information is an activity as old as oratory itself. Since the remote times of the sophists, ideas have been disseminated for persuasive and commercial purposes through it (ARAÙJO; BOCHNER; NASCIMENTO, 2012, p. 332).

Advertising began in Brazil in 1808. According to Abreu (2007, apud RAMALHO, 2008, p. 22):

The history of advertising in Brazil can be traced back to the arrival of the printing press in 1808, which enabled the first Brazilian newspaper, the Gazeta do Rio de Janeiro, to be published. The publication, which circulated only on Saturdays, already contained advertisements for houses, books and slaves.

The Vargas government approved and authorized radio broadcasting. Ràdio Nacional was one of the most renowned stations in 1936, and at this time it was already broadcasting commercial advertisements, as was the case with products from Laboratòrio Sidney Ross. They advertised Sonrisal, Colirio Moura Brasil, Urodonal and Elixir de Inhame (ARAÙJO; BOCHNER; NASCIMENTO, 2012).

Television came to Brazil during the 50s. According to Araújo et al. (2012, p. 334) "the 1950s imposed a new form of communication with the emergence of television, which would become the most popular media at the end of the 20th century". For the drug industries, the help of television made it easier to understand how to capture the information transmitted within reach of several people (RAMALHO, 2008).

Figure 1 shows one of the most viewed advertisements from the 1950s, which was the campaign for the drug Melhoral, which still circulates on the commercial pharmaceutical market. This drug is indicated as an analgesic and antipyretic in the treatment of flu and headaches (PROPAGANDAS HISTÓRICAS, 2016).

Figure 1- Drug advertisement from the 1950s

Source: Propagandas Históricas. (Available at: <http://www.propa gandashistoricas.com.br/2014/08/melhoral-alivio-imediato-anos- 50.html>. Accessed on: October 2, 2016).

In the 1960s, the King of soccer, Edson Arantes do Nascimento, better known as Pelé, lent his successful image to the Brazilian marketing of the medicine Biotômico Fontoura, illustrated in figure 2. This medicine is indicated for fortification as an appetite stimulant and also acts as an anti-anemic. The commercial was well accepted by Brazilian society at the time, and soon there was a publication by the author Monteiro Lobato, a friend of the medicine's creator, in which he decided to promote Biotômico Fontoura through the literary character Jeca Tatu, who at one point uses this drug to improve his state of health (PROPAGANDAS HISTÓRICAS, 2016).

Figure 2- Advertisement for the drug in the 1960s

Source: Propagandas Históricas. (Available at: <http://www.propa gandashistoricas.com.br/2014/12/biotomico-fontoura-pele-anos- 60.html>. Accessed on: October 2, 2016).

Today's advertising uses the most aesthetic and technological resources as a way of guaranteeing greater visual quality for the product. Unlike the advertisements of previous decades. However, the description of contraindications as well as information about side and adverse effects is continually missing from this advertising campaign.

Figure 3 shows the 2016 advertisement for the drug APRACUR, starring the country singer Leonardo and his son Zé Felipe with the slogan "from generation to generation". This drug is indicated for colds and flu. (PORTAL DA PROPAGANDA, 2016)

Figure 3 - Advertisement for the drug in 2016

Source: Portal da Propaganda. Available at: <http://www.portalda propaganda.com.br/portal/component/content/article/16-capa/46054- comunicafilmes-reune-leonardo-e-seu-filho-ze-felipe-pela- primeira vez>. Accessed on: October 2, 2016.

The consumption of medicines is related to advertising and marketing, which are widely used in the various media (KOTLER; ARMSTRONG, 2005). That's why many people find information through the media. Anvisa (2016) says that during the campaign there is an incentive to continue taking the medication.

In today's consumer society, medicine is conceived as a product that needs to be constantly updated and renewed in its presentation. This is coupled with science, which aims to guarantee the efficacy and safety of the product for the user. In addition, the symbolism of health strengthens consumer habits by presenting medicine to the consumer in a seductive and saleable way, such as "immediate pain relief", "improves your physical performance", "increases your appetite", "makes you calm", and not with the simple recovery and quality of health (ARANDA DA SILVA, 2007).

Today, the internet is the greatest media invention, used by a large number of people. Through this means of communication, there are ways of putting the commercial marketing of drug companies into practice. Therefore, it can be inferred that self-medication is related to the ease of obtaining information through this media (SOUZA;

MARINHO; GUILAM, 2008).

The pharmaceutical industry can be described as a set of multi-product oligopolies, differentiated into segments of specific therapeutic classes, whose consumption is strongly mediated by the need for medical prescription. Products are differentiated by therapeutic class, active ingredient, chemical composition and packaging, generating a wide range of products for the consumer (OLIVEIRA; LABRA; BERMUDEZ, 2006).

The aim of drug advertisements is to influence the public through actions that promote and/or induce the prescription, dispensation, purchase and use of drugs. In the majority of cases, drug advertisements use professionals who are able to persuade the public, promoting only their benefits and often omitting information related to the safety of the drug (SOARES, 2008).

Among the factors pointed out as influencing self-medication behavior, the marketing of pharmaceutical companies stands out as one of the main causes. Other reasons include the slowness, dissatisfaction and poor quality of public health services, the large amount of medical information available on the internet, and easy access to medicines, partly because the private sector is the main distributor of medicines to the population (AQUINO et al., 2008; NASCIMENTO, 2005).

By analyzing one hundred pieces of advertising for medicines and comparing the content of these ads (images, text and therapeutic indications for each product) with the provisions required by the legislation that regulates the practice of pharmaceutical advertising, researchers Nascimento and Sayd (2005) concluded that all one hundred pieces infringe at least one article of the standard, with the average per piece exceeding four infringements.

The media has the power to influence public opinion and contributes to the process of educating society. Radio plays an important role in this process and the broadcaster has the function of being close to the listener, transmitting information through their broadcasts. (BRASIL, 2008).

There is a significant weakness in the model for regulating drug advertising in Brazil (NASCIMENTO; SAYD, 2005; SOARES, 2008).

This is highlighted by reports of overpricing of raw materials and abuses in the price of medicines in recent years, as well as market failures. The state has a regulatory role vis-à-vis the industrial sector.

A regulatory structure is needed to mitigate this situation. Among the main characteristics that would explain state intervention are the following: concentrated markets, high barriers to entry, inelastic demand, variations in product prices and asymmetry of information (OLIVEIRA; LABRA; BERMUDEZ, 2006).

1.5 Medicines legislation

In RDC 102/2000, the main contraindication of the drug must be clearly and precisely stated in Portuguese. This rule prohibits advertisements containing comparisons that are not based on information proven by clinical studies published in indexed publications and the practice of provoking fear, anguish or suggesting that a person's health will or could be affected by not using the medicine. It prohibits the attribution of curative properties to the medicine when it is intended only for symptomatic treatment and the control of chronic diseases and states that advertising cannot suggest the absence of side or adverse effects, or use expressions such as: "innocuous", "safe" or "natural product" (BRASIL, 2000).

In a nutshell, it says that prescribing medicines, magistral and/or officinal preparations and other health products in Brazil is only permitted to legally qualified professionals in accordance with specific laws, as follows: Doctors, who have the natural right to prescribe after diagnosis; Dental surgeons, who can only prescribe for dental use - Law 5081/66; Veterinary doctors, who can only prescribe for veterinary use - Law 5517/68; Nutritionists cannot prescribe medicines. Law 8234/91 only allows these professionals to prescribe nutritional supplements, and also defines the conditions under which this can happen (CRFPA, 2016).

A provisional measure authorizing the sale of non-prescription medicines (PIMs) in supermarkets, warehouses, emporiums and convenience stores was approved by Congress, but later vetoed in May 2012, on the grounds of increased self-medication, and it is well known that in some regions of Brazil this rule is not observed, and it is possible to find

medicines even at gas stations. ANVISA authorized Mips to remain on the shelves by means of RDC 41/2012, because there has been no reduction in the number of poisonings.

According to ABIMIP (2016), over-the-counter medicines were first reported in Brazilian health legislation, Law No. 5.991 of December 17, 1973, which deals with the health control of medicines. In 2003, RDC No. 138 of May 29 was published, regulating the Agency's PIMs. In 2016, Anvisa presented a new regulation for PIMs, which was RDC No.

98, of August 1, which talks about the current criteria for classifying a medicine, such as: time on the market without a prescription, safety of the drug; symptomatology, its use for a short period of time, if it is manageable by the patient and if it presents a low risk potential, in addition to necessarily not causing dependence.

In Brazil, the laws governing the activities of these establishments had an important milestone in 1973, when Federal Law No. 5.991 was created, which sets out the sanitary control of the trade in drugs and medicines, pharmaceutical supplies and others. Under this law, places where medicines are sold (pharmacies and drugstores) are included in the list of controlled places, where relevant factors are sought to meet the changes that have taken place in this area, such as aspects of professional pharmaceutical practice (ROMANO-LIEBER; CUNHA; RIBEIRO, 2008).

One of the functions of health surveillance is to monitor and evaluate the use of medicines, with a view to obtaining concrete results and improving society's quality of life.

Health surveillance is the oldest public health activity. Since ancient times, social organizations have tried to control the key points of collective life and the threats to health and life itself (COSTA, 2000).

In Brazil, Health Surveillance became fundamental in the late 1980s to early 1990s, when the new Constitution of the Federative Republic of Brazil was promulgated in 1988, introducing a new concept and a new breadth of relations in the area of health, proving that health is a right for all and a duty of the State to provide it (SOUZA; STEIN, 2007).

3 METHODOLOGY

3.1 Nature of the study

The research was carried out through a simple descriptive, cross-sectional observational investigation carried out in the Cangalheiro, Salobro and Seriema neighborhoods of the city of Caxias-MA during the months of September to November 2016. A mixed questionnaire with twenty-six open and closed questions was used as the study instrument.

3.2 Place of research

The survey was carried out in the Cangalheiros, Salobro and Seriema neighborhoods in the city of Caxia-MA. The aim was to choose neighborhoods that were geographically distant and had different economic and social characteristics.

Caxias is located in the mesoregion of Eastern Maranhão, with 5,313.2 km^2 , bordered to the north by the municipalities of Codó, Aldeias Altas and Coelho Neto; to the south by the municipalities of Sao Joao do Sóter, Parnarama, Matoes and Timon; to the east by the state of Piaui and to the west by the municipalities of Codó and Sao Joao do Sóter and Gonçalves Dias (IBGE, 2008).

The Cangalheiro neighborhood is located in the southern part of the municipality of Caxias-MA, on the right bank of the Itapecuru River. It is bordered to the north by Os Três Coraçoes, to the south by the Volta Redonda neighborhood, to the east by Morro do Alecrim and to the west by the Itapecuru River.

Salobro is a street near the Campo de Belém and Trezidela neighborhoods. It lies on the left bank of the Itapecuru River in the southeast of the municipality of Caxias.

The other neighborhood surveyed, Seriema, is made up of residential houses and housing estates. This neighborhood has a higher economic level than the other neighborhoods. According to some residents, the name of the neighborhood is related to the fact that in the brief past there was a large concentration of birds called Seriema.

As for the number of inhabitants and houses in the neighborhoods surveyed, the Cangalheiro neighborhood has an estimated 1,401 houses and a total of 5,871 inhabitants. The Salobro neighborhood has an estimated 815 houses and 3845 residents. The Seriema neighborhood has an estimated 2637 homes and 9614 inhabitants (PROGRAMA DE COMBATE A DENGUE, 2016).

3.3 Participants and information gathering

Data was collected by visiting the homes of neighborhood residents and applying a questionnaire designed by the researcher and author of the study. The survey involved a population of 900 people, divided into 300 respondents per neighborhood. The sample was obtained through systematic random sampling. The inclusion criteria for taking part in the survey were: age 18 or over and belonging to one of these neighborhoods.

3.4 Data analysis

After applying the questionnaire, the data was analyzed using Excel 2010 and the SPPS 20.0 *software,* which made it possible to present the results in graphs and tables.

The absolute and relative frequencies of the graphs were analyzed by comparing the variables presented between the neighborhoods according to the questions in the questionnaire and the research objective.

3.5 Risks and benefits

The study offers minimal risk to the participant, while the chance of direct and indirect benefit is quite significant, since by discussing self-medication, it provides both the interviewer and the interviewee with clarification about the real situation of public health and the risks that self-medication can manifest. We made sure to respect the individuality of each participant, thus avoiding any kind of discomfort during the questionnaire. It should be noted that the free and informed consent form (FICF) was presented and explained to guarantee the confidentiality of information.

3.6 Ethical and legal aspects

In order to carry out this research, authorization was sought from the people involved by means of a statement. Once this has been done, the work is in the process of being submitted to the Research Ethics Committee to certify compliance with all the ethical aspects of scientific research with human beings contained in Resolution No. 466/2012 of the National Health Council. With research ethics committee opinion number 2772. Also in accordance with the same Resolution, which in its second chapter deals with terms and definitions, which were presented through the Informed Consent Form (ICF).

4   Results and discussion

The sample analyzed in this study comprised a total of 300 people per neighborhood, with a total sample of 900 people. It can be seen from the results in Table 1 that the predominant age group in the three neighborhoods analyzed were people over 40, and that there was also a higher absolute and relative percentage of female interviewees. A survey of elderly people from all income brackets on self-medication found proportions of 17% in Bambui[6] , MG and up to 60% in Salgueiro, PE[7] .

These results should be clarified, because even though men consume drugs, the incidence of women buying them is higher (BEZERRA, 2006). Lessa and Bochner (2008) state that women show their feelings more often, have a higher percentage of depression and hospital admissions than men.

Table 1 - Epidemiological and sociodemographic profile of the Cangalheiro, Salobro and Seriema neighborhoods

	Sociodemographic characteristics	Residents	
		n (300)	%
	Age group		
	18-25 years	90	30
	26-30 years	29	13
	31-40 years	54	18
Kangaroo	Over 40 years old	117	39
	Sex		
	Male	132	44
	Female	168	56
	Sociodemographic characteristics	Residents	
		n (300)	%
	Age group		
	18-25 years	60	20
	26-30 years	60	20
Salobro	31-40 years	84	28

---

6 - Loyola Filho et al. (2005).
7 - Sa et al. (2007).

Sociodemographic characteristics	Residents	
	n (300)	%
Over 40 years old	96	32
Sex		
Male	66	22
Female	234	78

	Sociodemographic characteristics	Residents	
		n (300)	%
	Age group		
	18-25 years	54	18
Seriema	26-30 years	69	23
	31-40 years	96	32
	Over 40 years old	81	27
	Sex		
	Male	105	35
	Female	195	65

Source: Ana Florise Morais Oliveira, Field research, 2016.

Graph 1 shows the percentage distribution of residents of the Cangalheiro, Salobro and Seriema neighborhoods who are aware of the risks of self-medication. Percentage results were obtained: Cangalheiro (28%), Salobro (34%) and Seriema (38%) of people who have information about the risks of self-medication.

From the results of the graph, it can be inferred that the neighborhood with the most information about the risks of self-medication is the Seriema neighborhood. Self-medication causes risks and is related to the lack of information about the unwanted effects and adverse reactions that can be caused by drugs (MARQUES, 2006).

Graph 1 - Percentage distribution of residents of the Cangalheiro, Salobro and Seriema neighborhoods who are aware of the risks of self-medication.

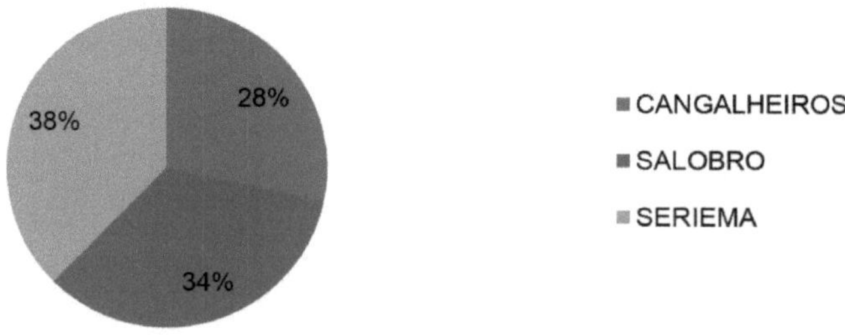

Source: Ana Florise Morais Oliveira, field research, 2016.

Graph 2 shows the percentage distribution of drugs purchased with a doctor's prescription among residents of the Cangalheiro, Salobro and Seriema neighborhoods.

Of the 300 people interviewed in the Cangalheiro neighborhood (27.3%) said that they used a doctor's prescription before buying medicine. In the Salobro neighborhood (25.7%) and in the neighborhood (50.3%).

After the analysis, it is possible to state that the Seriema neighborhood has the smallest discrepancy in terms of the evaluation criteria between those who answered yes or no. 50.3% of those interviewed in Seriema use a doctor's prescription before taking pharmacotherapy. This may be due to the high economic prevalence among these residents, which makes it easier for them to see a doctor. According to Nascimento et al. (2005), access to quality healthcare in Brazil is not a reality for the majority of the population, since in order to obtain it they need to be able to afford a private healthcare plan.

Graph 2- Percentage distribution of prescription drugs among residents of the Cangalheiro, Salobro and Seriema neighborhoods

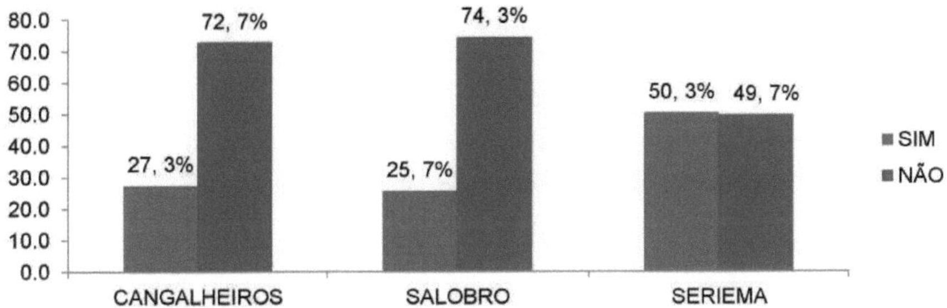

Source: Ana Florise Morais Oliveira, field research, 2016.

Graph 3 shows the percentage distribution of the last three months of possible symptoms among those interviewed in the survey. The percentage results were close between the Seriema and Salobro neighborhoods, with an average of 60% of those interviewed not showing any symptoms of illness in the last three months. In the Cangalheiro neighborhood, 54.7% answered that they had not shown any signs of illness either.

Graph 3   - Percentage distribution of the last three months among residents who presented symptoms in the Cangalheiro, Salobro and Seriema neighborhoods.

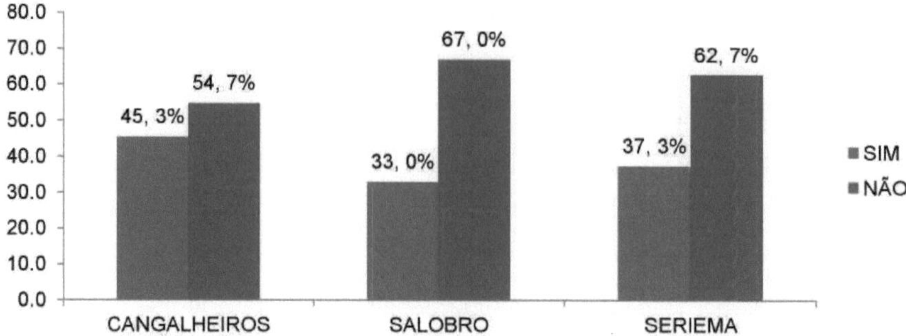

Source: Ana Florise Morais Oliveira, field research, 2016.

Graph 4 shows the distribution of the drugs most commonly used by respondents in the neighborhoods surveyed on a routine basis. It can be seen that the most commonly used drugs in the three neighborhoods are painkillers, anti-inflammatory drugs and antifungal drugs. This coincides with the studies by Galato, Madalena and Pereira (2012), which showed a 52.3% rate of use of painkillers among the survey respondents.

Similar results were found regarding the predominance of analgesics as the class of medication most used by the population, in the studies by Freitas; Martins (2008) and Arrais et al. (1997), even with the significant difference of thirteen years, the research carried out by Lalama (1999), which describes anti-inflammatories, analgesics and antibiotics as the most consumed in Brazil, is still valid.

Graph 4 - Percentage distribution of the classes of medicines used by residents of the Cangalheiro, Salobro and Seriema neighborhoods in their daily lives.

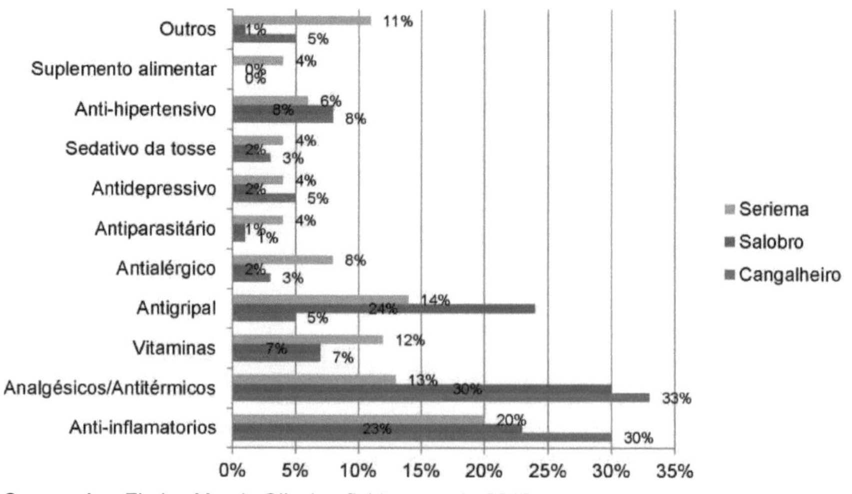

Source: Ana Florise Morais Oliveira, field research, 2016.

Graph 5 shows the percentage distribution of symptoms for the use of a particular drug in the three neighborhoods surveyed. There was a predominance among the interviewees of headache as the most recurrent symptom for the use of a drug.

Ogawa et al. (2008) state that the population needs to be aware of the risks of self-medication. The main reason for self-medication is headache and it can be a sign or symptom related to another health problem which may be more serious and require specific treatment and care.

Graph 5 - Percentage distribution of the symptoms presented for the use of a given medication in the residents of Bairro Cangalheiro, Salobro and Seriema.

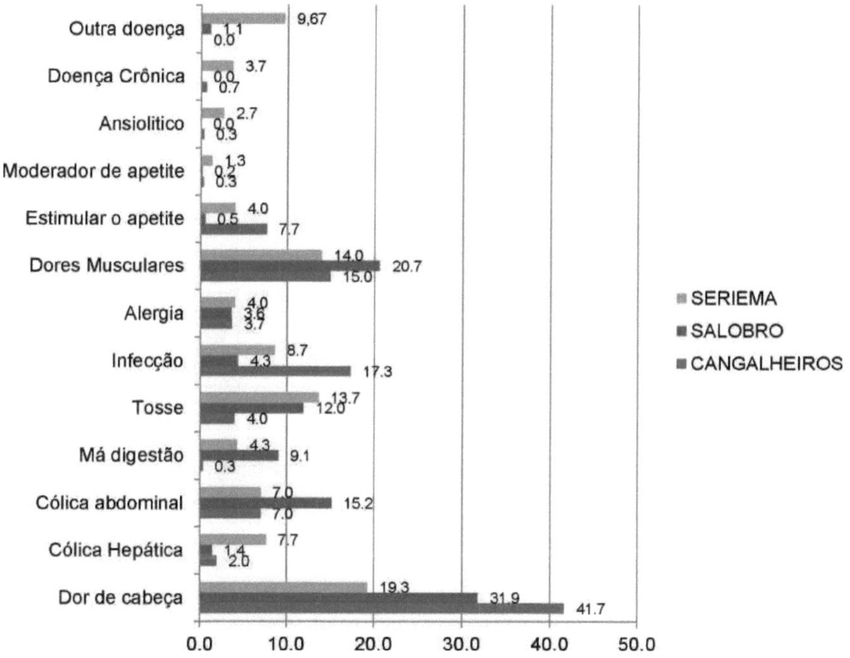

Source: Ana Florise Morais Oliveira, field research, 2016.

According to Graph 6, more than half of the participants in the survey were unaware of the side effects caused by the use of medicines. Of the 300 people interviewed in the Seriema neighborhood, 65.3% were unaware of the side effects of the drugs they use for therapeutic treatment.

Side effects can be mild or severe, caused by drug interactions which can lead to greater health risks (FUCHS et al., 2006).

According to Katzung (2005), information on when and how to administer the medication and its therapeutic duration is the sole responsibility of the doctor and pharmacist for each patient. In addition, the indication that was prescribed, the name of the drug and the duration of therapy should be recorded on each label, preventing easy identification when an overdose occurs.

Graph 6   - Percentage distribution of knowledge of the side effects of medicines used by residents of Bairro Cangalheiro, Salobro and Seriema

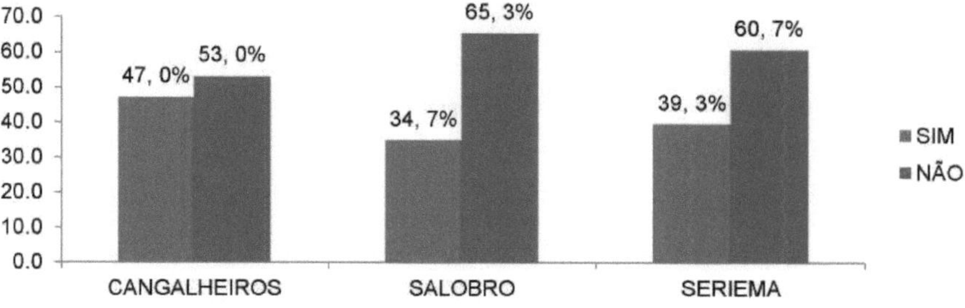

Source: Ana Florise Morais Oliveira, field research, 2016.

Graph 7 shows the percentage of self-medication among the total survey sample as a way of verifying the real situation of self-medication among residents of the city of Caxias-MA. It can be seen that 74.8% of the 900 interviewees use self-medication to treat illnesses. This figure shows that the population in this study needs guidance when it comes to the risks of poisoning as well as the aggravation of illnesses caused by the improper use of medicines.

The majority of the Brazilian population has difficulty in accessing public health care, where part of society in the poverty bracket cannot afford a health plan, and in these conditions the practice of self-medication is common. However, the financial criterion is not enough to explain the number of people who use this method, other factors such as social class, level of education, access to information about medicines, and especially the cultural factor clarify the reasons for the practice of self-medication (NASCIMENTo, 2005).

Graph 7 - Percentage distribution of self-medication in the total sample

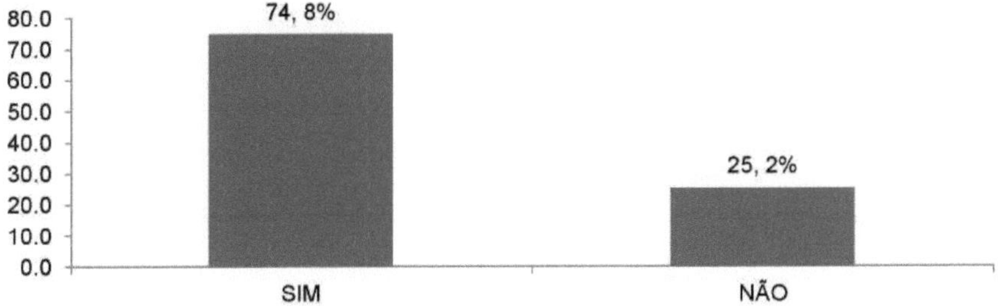

Source: Ana Florise Morais oliveira, field research, 2016.

Graph 8 shows the high percentage of interviewees who use television as their main means of obtaining information. Since it is the communication vehicle most used by Brazilian society today.

Magazines, billboards and television advertisements are places where information is displayed, where these media induce the consumption of medicines, using phrases that captivate the consumer, artistic personalities, and miraculous proposals. (MUSIAL, et al., 2007)

Graph 8 - Percentage distribution of residents in the Cangalheiro, Salobro and Seriema neighborhoods regarding the use of the media

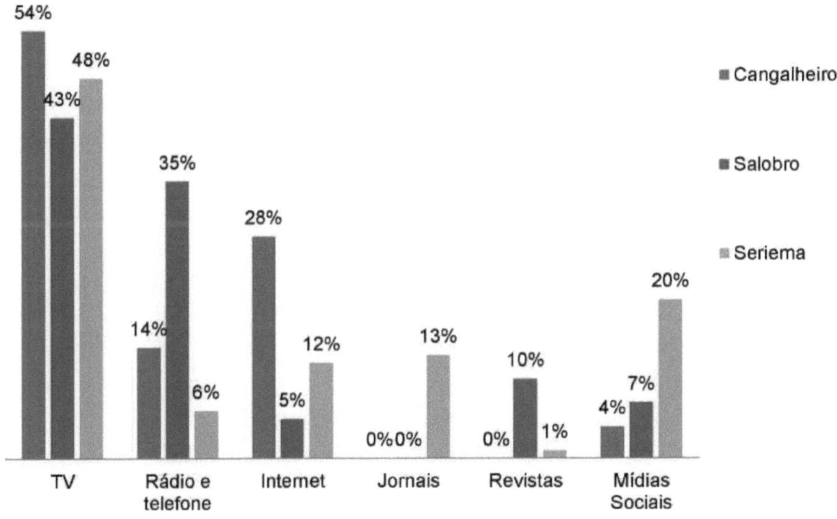

Source: Field research, 2016.

Graph 9 shows the predominance of results among the participants from the three neighborhoods surveyed with regard to the influencing factors that determined the practice of self-medication, with the following results: the reuse of the same medicine because it had been administered other times and nothing had happened to compromise the individual's health. Another important factor is the sharing of information, as more than 30% of the Cangalheiro neighborhood responded that they had received advice from someone else, be it a friend or family member, about the use of a particular medication.

Graph 9 - Percentage distribution of the reasons for self-medication in the Cangalheiro, Salobro and Seriema neighborhoods.

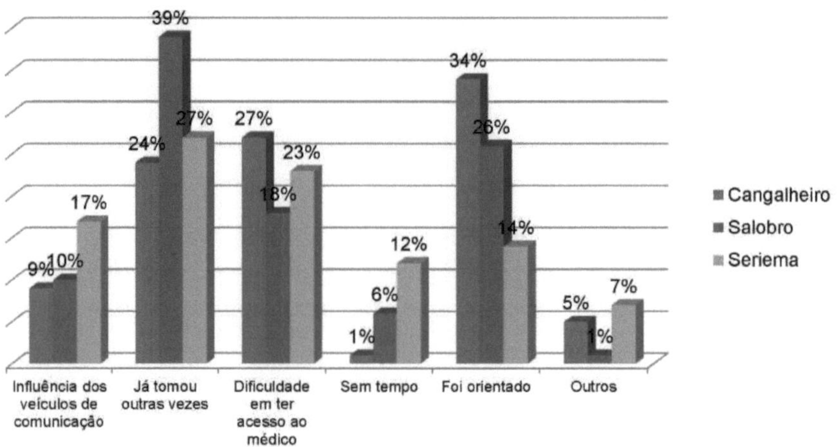

Source: Ana Florise Morais Oliveira, field research, 2016.

Graph 10 shows the distribution of self-medication in the neighborhoods surveyed, in terms of those who said that they use it and those who don't. It can be said that the interviewees from the three neighborhoods had a high number of people who self-medicate, but the Seriema neighborhood stands out in percentage terms compared to the other two neighborhoods, as well as in the discrepancy between those who answered yes (28%) and no (5%).

Graph 10 - Percentage distribution of residents of the Cangalheiro, Salobro and Seriema neighborhoods who self-medicate.

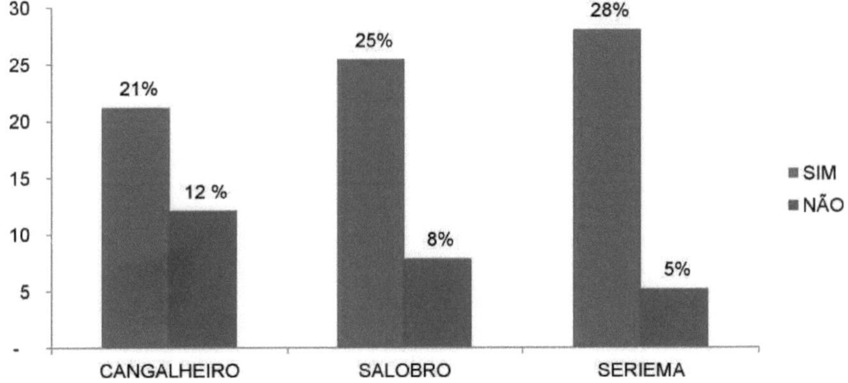

Source: Ana Florise Morais Oliveira, field research, 2016.

Table 2 describes the Chi-squared test or Fischer's exact test to assess the influence of the self-medication profile among residents of the Cangalheiro, Salobro and Seriema neighborhoods in the city of Caxias, MA, with $p < 0.05$ adopted as significance. SPPS 20.0 software was used to analyze the relative and absolute frequency of the data collected. The data analysis methods were descriptive statistics with measures of central tendency and dispersion for continuous variables and percentages and numbers for categorical variables. When necessary, numerical variables were categorized by median. The chi-square test (X^2) was used to test the homogeneity of the proportions. A significance level of 95% was set. The prevalence of self-medication was estimated using the 95% confidence interval (95% CI).

The calculated **p-value** is equal to 0.05, so there is a direct relationship of dependence between the resident population and self-medication.

We can also see that the percentage discrepancy in the Seriema neighborhood with regard to yes and no answers in the cross-tabulation is more pronounced than in the other neighborhoods.

Table 2 - Chi-squared test

	Value	df	Sig. Assint. (2 sides)
Pearson's Chi-squared	34, 546[a]	2	0, 000
N of Valid Cases	900		

For α=0, 05

BAIRRO * AUTOMEDICATION Cross-tabulation

Counting

| | | AUTOMEDICATION | | Total |
		YES	NO	
	KANGAROO	191	109	300
BAIRRO	SALOBRO	229	71	300
	SERIEMA	253	47	300
Total		673	227	900

Source: Field research, 2016

5   CONCLUSION AND PROPOSAL

Self-medication is a reality in Brazil and around the world, but you need to be aware of its consequences and the risks involved.

This research aimed to analyze the self-medication profile of residents of the Cangalheiro, Salobro and Seriema neighborhoods in the city of Caxias-MA between September and November 2016. The results obtained from the survey showed that self-medication is a common habit and a worrying reality in today's Brazilian society.

According to the collection of bibliographic information and the results of the questionnaire applied in the neighborhoods, it can be deduced that there are several causes that led the interviewees to make the decision to self-medicate: the difficulty of the population's financial resources, the deficiency of the public health service, the sharing of information and/or guidance from other people who don't have the appropriate training for the task, the media which broadcast advertising campaigns for medicines without transparency and often omitting the side-effects of the products which are marketed. In addition to these factors, it is important to state that the problem of self-medication is also a cultural one, as many people are unaware of the risks and dangers that this activity can have on human health, yet they continue to practice it.

It was also found that there is a need for structural changes at the level of the drug sales system, as most of the drugs used by the interviewees were obtained without a medical or dental prescription.

As for the symptoms reported by the survey participants, headaches, muscle pain and infection related to the microorganisms stand out.

It is believed that this work will contribute to the scientific community of the Mauricio de Nassau College as a source of research as well as to other scholars and to reinforce how necessary it is to discuss this issue.

It's worth remembering that clarifying adverse reactions and side effects makes all the difference to the effectiveness and efficiency of the drug, thus avoiding rampant consumption and the myth of the "miracle cure".

It is proposed that measures be adopted as soon as possible through educational campaigns in visible places such as public squares, schools and colleges. It is up to health professionals, especially pharmacists, to raise awareness in the community.

6 REFERENCES

SANITARY VIGILANCE AGENCY - ANVISA. **Reaçôes adversas a medicamentos** . Available at: < http://portal.anvisa.gov.br/documents/33868/ 2894427/Rea%C3%A7%C3%B5es+Adversas+a+Medicamentos/1041b8af-9cde-4e94- 8f5c-9a5fe95f804d >. Access em: 9 out. 2016.

. **Conceitos gérais sobre medicamentos** . Available at: < http://www. anvisa.gov.br/hotsite/genericos/profissionais/conceitos.htm#9 >. Access em: 9 out. 2016.

. **Propaganda:** Brasilia. Available at: < www.anvisa.gov.br/propaganda/ folder/uso indiscriminado.pdf >. Accessed: 15 Aug. 2016.

ALBALADEJO, F. M., DiEZ, BJ Aspectos sociológicos del empleo de medicamentos In: ALBALADEJO FM, DiEZ, BJ, **Principios de Pharmacologia Clinica.** Barcelona. v. unique. pp. 271-281.2002.

ALMEIDA, P.; SILVA, D. Anti-inflammatórios não esteroidais mais dispensados em uma farmâcia de manipulaçâo do municipio de Itaperuna-Rio de Janeiro, Brasil. **Acta biomédica brasiliensia** , Rio de Janeiro, v. 4, n. 1, p. 24-35, 2013.

ALONZO, HGA; CORRÊA, C. L. Analgésicos, antipiréticos e antiinflamatórios não esteroidais: dados epidemiológicos em seis centros de controle de intoxicaçôes do Brasil. **rev. BrasToxicol.** , v. 14, n. 2, p. 49-54, 2001.

ANGELL, Marcia. **The truth about pharmaceutical laboratories** . Rio de Janeiro. Record, 2007.

AQUINO, DS; BARROS, JAC; SILVA, MDP Automedicaçâo e os acadêmicos da área de saùde. **Ciência & Saùde Coletiva** , v. 15, n. 5, p. 2533-2538, 2010.

ARANDA DA SILVA, JA **Existe uma ligaçâo directa entre a qualidade de vida ea automedicaçâo** . Interview dismissed PRISFAR News. Available at: < http://www.prisfar.pt/news/news-n9-f.asp >. Access em: 15 Aug. 2016.

ARAÙJO, CP; BOCHNER, R; NASCIMENTO, AC Marcos legais da propaganda de medicamentos: avances e retrocessos. **Physis** , v. 22, p. 331-346, 2012.

ARAÙJO, Carolina Pires. **Drug propaganda : give persuasive strategies to embate discursively** . 2012. Dissertação (Mestrado) - Programa de Pôs-Graduaçâo em Informaçâo, Comunicaçâo em Saùde, Instituto de Comunicaçâo e Informaçâo Cientifica e Tecnològica em Saùde, Fundaçâo Oswaldo Cruz, Rio de Janeiro, 2012.

ARRAIAIS, PS D; COELHO, HLL; BATISTA, MCDS; CARVALHO, ML; RIGHI, R.E .; ARNAU, JM **Perfil da Automedicaçâo no Brasil** . Saùde pùblica. v.31, p.71-79, 2005.

AZEVEDO, FA A Toxicology is the Future. **rev. Inter** . v.3, n.3, 2010

BEZERRA, Janielde Lopes. **Automedicaçâo por usuàrios de uma commercial pharmacy no Bairro Pirajà na cidade de Juazeiro do Norte-CE.** v. 69. Curso de especializaçâo em assistencia farmacèutica. Escola de Saùde Pùblica do Cearà. 2006

BRICKS, Lucia Ferro. **Judicious use of medicines in children. J. Pediatrician** . Porto Alegre, v. 79, suppl. 1, 2003.

BEZERRA, Janielde Lopes. **Automedicaçâo por usuàrios de uma famàcia comercial no Bairro Pirajà na cidade de Juazeiro do Norte-CE** . 2006, v. 69.

Curso de especializaçao em assistance pharmacèutica. (Escola de Saùde Pùblica do Cearà. Juazeiro do NORTE-CE)

BOCHNE, R. R. Papel da Vigilância Sanitària na prevençâo de intoxicaçôes na infância, **Revisa** , n. 1, v. 1, p. 50-57, 2005.

BORTOLON, PC et al. **Anàlise do perfil de automedicaçâo em mulheres idosas brasileiras** . Ciència & Saùde Coletiva, v.13, n.4, p.1219-l226, 2008.

BRAZIL. Agència Nacional de Vigilância Sanitària. **Ocontrole necessàrio para as propagandas na construçâo da cidadania** . Available at: < http://www.anvisa.gov. br/divulga/noticias/2005/261205_1_texto_de_esclarecimento.pdf >. Access em: 6 Aug. of 2016.

. Agència Nacional de Vigilância Sanitària. **Gerência de Monitoramento e Fiscalizaçâo de Propaganda, Publicidade, Promoçâo e Informaçâo de Produtos sujetiosa Vigilância Sanitària** (GPROP/DiFRa).

BUCARETCHI, F. **Self-medication for children** . Diàrio do Noroeste, Paranavai, 2007.

CANTARINO, A. Marketing x Pharmaceutical Legislation. **2° Congresso Cientifico da UniverCidad** e - Resumo expandido, Rio de Janeiro, 2007. Available at: < http://www.univercidade.br/pesqcient/pdf/2007/amb_mkt.pdf >. Access em: 14 Aug. 2016.

CARVALHO, M. F.; PASCOM, A. R. PP; SOUZA-JÙNIOR, P. R. B.; DAMACENA, GN; SZWARCWALD, CL. Utilization of medicines by the Brazilian population, 2003. **Cad. Saùde Pùblica** , vol. 21, suppl.1, p.100-108, 2005.

CASTRO, C. G. SO **Estudos de utilizaçao de medicamentos:** noçôes básicas Rio de

Janeiro, Editora FIOCRUZ, 2000.

CASSIANI, SHB **A segurança do paciente eo paradoxo no uso de medicamentos.** Revista Brasileira de Enfermagem. Brasilia, v. 58, n. 1, p. 95-99, 2005.

CENÇO, B. Automedicaçâo: isso tem que parar. **APM magazine (Associaçâo Paulista de Medicina)** . São Paulo, v. 610, p. 5-8, 2010.

CERVO, Amado Luiz; BERVIAN, Pedro Alcino. **Scientific methodology:** para uso for university students. 3rd ed. Sao Paulo: McGraw-Hill do Brasil, 1983.

CHACRA, N. A. B.; FAGUNDES, M. JD; PINTO, T. J. S. Marketing is drug promotion . In: STORPIRTIS, S. et al. **Clinical pharmacy is pharmaceutical care.** Rio de Janeiro: Guanabara Koogan, p.489, 2008.

CASSIANI, SHB A segurança do paciente eo paradoxo no uso de medicamentos. **Revista Brasileira de Enfermagem** . Brasilia, v. 58, n. 1, p. 95-99, 2005.

CHIAROTI, R.; REBELLO, NM; RESTINI, CDA A automedicaçao na cidade de Ribeirao Preto - SP eo papel de farmacèutico nessa pràtica. **Enciclopédia Biosphere, Centro Cientifico Conhecer** . Goiânia - GO, v. 6, n. 10, 2010.

CONSELHO REGIONAL DE FARMACIA DO ESTADO DE SP- CRFSP . Available at: <http://www. crfsp.or.br/index.História-da.../290-Surgimentodasboticas.http>. Accessed: 14 Nov. 2016.

. **História da Farmàcia** . Available at: < http://portal.crfsp.org.br/index.php/ historia-da-farmacia-.html >. Access em: 20 set. 2016.

CONSELHO REGIONAL DE PHARMACIA DO ESTADO DO PA - CRF-PA .

Available at: <http://www. crfpa.org.br/sitescd/crfpa/>. Access em: 26 Nov. 2016.

CRUZ, PS; CARAMONA, M.; GUERREIRO, MP Uma reflexao sobre a automedicaçao e medicamentos nao sujeitos a receita médica. **Revista Portuguesa Farmacoterapia** , v.7, p.83-90, 2015.

DAMASCENO, D. D.; TERRA, FS; ZANETTI, HV; D'ANDREIA, E. D.; SILVA, HL R.; LEITE, J. A.; Automedicaçâo entre graduandos d enfermagem, farmàcia e odontologia da universidade federal de Alfenas. **rev. Min. Enferm** ., v. 11, n. 1, p. 48-207, 2007.

DE OLIVEIRA, Egléubia Andrade; LABRA, Maria Eliana; BERMUDEZ, Jorge. A produçâo pùblica de medicamentos no Brasil: uma visâo geral. **I'm falling. Saùde Pùblica** , Rio de Janeiro, v. 22, n. 11, nov. 2006.

DIAS, José Pedro Sousa. **A Farmàcia ea História Uma introduçâo à História da Farmàcia, da Pharmacologia e da Terapèutica** . Available at: file:///C:/Users/Admin/Downloads/Farmacia-e-Historia[1].pdf . Acesso em: 27 sets. of 2016.

FARIA, FAC; MARZOLA, C. Farmacologia dos anestésicos locais - consideraçôes gerais. **BCI** , Curitiba, v. 8, n. 29, p. 19-30, 2001.

FERNANDES, L.Ç.; TORRES, IL; KAUFFMANN, C.; CASTRO, L.C. et al. **Anàlise da asistencia Farmacèutica no SUS no Vale do Taquari - RS** . Relatório Final de pesquisa apresentado ao Centro universitários UNIVATES, 2008.

FERREIRA, Vitor F.; PINTO, Ângelo C. A Fitoterapia no Mundo Atual. **Quim. New** ,

Vol. 33, No. 9, 1829, 2010

FIGUEIRAS, Maria Joâo . et al. Medicamentos Genericos: common beliefs of the Portuguese population . **rev. I'm wearing. Clin. Geral,** v. 23, p. 43-51, 2007.

FREITAS, EO; MARTINS. **Concepçôes de saùde no livro didâtico de ciências.** Ensaio Pesquisa em Educaçao em Ciências, 2008.

FUCHS et al. **Clinical pharmacology:** fundamentals of rational therapeutics . 3rd ed. Rio de Janeiro: Guanabara Koogan, 2006.

FUNDAÇÂO OSWALDO CRUZ- FIOCRUZ. **Sistema Nacional de Informaçôes Tóxico-Pharmacológicas** . Available at: < http://www.fiocruz.br/sinitox/2003/ umanalise2003.htm >. Acesso em: 19 sets. of 2016.

GALATO, D.; MALADENA, J.; PEREIRA, G. Automedicaçâo em estudes universitârios: a influence da área de formationaçâo. **Ciência e Saùde collectiva** , v.17, p. 3323-3330, 2012.

GANDOLFI, E.; ANDRADE, MGG Eventos toxicológicos relacionados a medicamentos no Estado de Sâo Paulo. **rev. Saùde Pùblica** , v. 40, n.6, p.56-64, 2006.

GELLER, M.; KRYMCHANTOWSKI, AV; STEINBRUCH, M.; CUNHA, KS;

RIBEIRO, MG; OLIVEIRA, L.; OZERI, D.; DAHER, JPL Utilizaçâo do diclofenaco in clinical practice : review das evidences terapêuticas e açôes pharmacológicas. **rev. Arm. Clin. Med.,** Sâo Paulo, v.10, n. 1, p. 29-38, Jan./Feb., 2012.

GOODMAN, Louis S.; GILMAN, Alfred (Ed.). **The pharmacological basis of therapeutics: a textbook of pharmacology, toxicology and therapeutics for**

physicians and medical students . 3rd ed. New york. Macmillan. 1965.

GUYTON, AC; HALL, JE **Treatise on Medical Physiology** . 10. ed. Rio de Janeiro: Guanabara Koogan, 2002.

HUGHES, CM; MCELNAY, JC; Fleming, GFBenefits and risks of self medication. Drug Safety. **An International Journal of Medical Toxicology and Drug Experience,** v. 24, p.1027-1037.2001.

INSTITUTO BRASILEIRO OF GEOGRAPHY AND STATISTICS - **IBGE** . Available at: < http://www.cidades.ibge.gov.br/xtras/perfil. php?lang=&codemun=210300 >. Access em: 16 out. 2016.

BRAZILIAN CONSUMER DEFENSE INSTITUTE - IDEC. Venda SEM prescriçâo. **Magazine do Idec.** Available at: <http://www.idec.org.br/uploads/ revistas_materias/pdfs/2004-06-ed144-capa-remedios.pdf> . Acesso em: 20 de ago. of 2016.

INSTITUTO VIRTUAL DOS FÀRMACOS DO RIO DE JANEIRO - **IVFRJ** . Available at: < www.ivrj.ccdecania.ufrj.br >. Acesso em: 21 sets. of 2016.

JESUS, P. R. C. **A automedicaçâo no Brasil** : um sintoma a ser analysado.

Available at: <http://cafehipócrates.org/2007/11/07/a-automedicacao-no-brasil-um-sintoma-a-ser-analisado>. Access em: 10 de ago. of 2016.

KATZUNG, BG Basic **and clinical pharmacology** . 9th ed. Rio de Janeiro: Guanabara Koogan, 2005.

KOTLER, P; ARMSTRONG, G. **Principles of Marketing** . 9th ed. Sao Paulo: Pearson/Prentice Hall, 2005.

LALAMA, M. **Perfil de consumo de medicamento en la ciudad de Quito** . Educate. med. contain, n.64, 7-9, 1999.

LAZZAROTTO, E. M. **Gestâo de serviçöes de saùde:** condiçöes de trabalho nas organizaçôes. Rattlesnake: Saber Column, 2004.

LEAPE, LL. et al.. Systems analysis of adverse drug events. ADE Prevention Study Group. **JAMA,** v.1, pp. 35-43, 1995.

LEITE, SN; VIEIRA, M.; VEBER, AP Estudos de utilizaçao de medicamentos: Uma sintese de articles publicados no Brasil e na América Latina. **Ciência e Saùde Collective.** Rio de Janeiro, v.13, p. 793-802, 2008.

LIMA, H.; FILHO, M. Anti-inflammatórios nao-esteroides eo uso indiscriminado: Um estudo em drogarias no municipio de Pimenta Bueno-RO. **UningàReview,** v. 4, n., p. 13-20, 2010.

LIMA, GB; NUNES, LCC; BARROS, JAC Uso de medicamentos estadotos em domicilio em uma populaçâo atendida pelo programa saùde da familia. **Chat. collective health** , v.15, n.3, p. 3517-3522, 2010

LOURO, E. Liebor; ROMANO, NS; RIBEIRO, E. Adverse antibiotic events in patients of a university hospital. **rev. Saùde Pùblica** , v. 41, n. 16, p. 1042-1048, 2007.

LOYOLA, AIF; UCHOA, E.; FIRMO, THURSDAY; LIMA COSTA, MF Estudo de populional base sobre o consumo de medicamentos entre idosos: Projeto Bambui. **Cad Saude Publica** . v.21, p.545-553, 2005.

LOYOLA FILHO, A. I.; UCHOA E. Self-medication: motivation is characteristic of its practice. **rev. Med. Minas Gerais** . v. 12, p. 219-227, 2002.

MACIEL, MAM; PINTO, AC; VEIGA, VE; GRYNBERG, NF ECHEVARRIA,

A. Plantas Medicinais: a need for multidisciplinary studies . **Quimica Nova** , v.25, n.3, p.429-438, 2002.

MAGALHAES, SM S.; CARVALHO, W. S. Reaçôes adversas a medicamentos. In: GOMES, MJVM; REIS, AM M (Org.). **Pharmaceutical sciences:** an approach to hospital pharmacy. 1st ed. Sao Paulo: Athenaeum, 2001.

MARQUES, FB **Medicamentos e farmacêuticos** . Lisboa: Campo de Comunicaçao, 2006.

MENDES, Z.; MARTINS, AP; MIRANDA, AC; SOARES, MA; FERREIRA, AP; NOGUEIRA, A. Prevalência da automedicaçao da populaçao urbana Portuguesa.

Revista Brasileira de Ciências Farmacêuticas , v. 40, n. 1, p. 21-25, Jan./Mar., 2004.

MOLENTO, Marcelo Beltrao. Resistência parasitària em helmintos de equideos e propostos de manejo. **Ciência Rural** , Santa Maria, v.35, n.6, p.1469-1477, 2005.

MOTA, Daniel Marques; SILVA, Marcelo Gurgel Carlos yes; SUDO, Elisa Case; ORTUN, Vicente. Uso rational de medicamentos: uma anbordacion economica para tomada de decisôes. **Chat. saùde collective** . v.13, p.589-601, 2008.

NASCIMENTO, D. M. **Estudo do perfil da automedicaçâo nas differentes classes sociais na cidade de Anàpolis Goiàs** , **2005** . Available at: <http://www.prp.ueg. br/06v1/ctd/pesq/inic_cien/eventos/sic2005/arquivos/saude/estudo_perfil.pdf> .

Access em: 19 set. 2016.

NASCIMENTO, A. C.; SAYD, JD Ao persisterem os sintomas o medico deverà ser consultado: isto è regulaçâo? **rev. Saùde Coletiva** , v. 15, n. 2, p. 67-75, 2005.

NASCIMENTO, MC **Medicamentos** : menace ou apoio à saùde?. Rio de Janeiro, 2003.

NIES, Alan. S. **Principles of therapeutics** . In: HARDMAN, Joel G.; LIMBIRD, Lee E.; GILMAN, Alfred Goodman. Goodman and Gilman's: the pharmacological basis of therapeutics. 10th ed. New york. McGraw Hill..p.45-66, 2001.

OLIVEIRA, Eglèubia A.; LABRA, Maria E.; BERMUDEZ, Jorge. A produçâo pùblica de medicamentos no Brasil: uma visao geral. **I'm falling. public health** , Rio de Janeiro, v. 22, n.11, p.2379-2389, 2006.

OLIVEIRA, ALM; PELÓGIA, NCC Cephaleia como principal causa de automedicaçâo entre os professionnels da saùde nao prescritores. **rev. Dor** , Sao Paulo, v.12, n.2, 2011.

OLIVEIRA, Karla Renata de; MUNARETTO, Paula. Rational use of antibiotics: responsibility of prescribers, usuàrios and dispensadores. **CONTEXTO & SAÙDE IJUi MAGAZINE** . UNIJUi Publisher, v.9. n. 18. p.43-51, 2010.

PAN-AMERICAN DA SAÙDE ORGANIZATION - OPAS. **Medicine and technology.** Available at: < http://www.opas.org.br/ Medicamentos/ temas.cfm?id =46&CodBar ra=1 >. Acesso em: 20 de ago. of 2016.

PEPE, VLE; OSORIO-DE-CASTRO, CGS Prescriçao de Medicamentos. In: **Formulàrio Terapèutico Nacional** . 1.ed. Brasilia DF: Ministerio da Saùde, 2008, p.1- PEREIRA, JR; SOARES, L.; HOEPFNER, L.; KARUGER, KE; GUTTERVILLE, M.

IT.; TONINI, KC; DEVEGILI, YES; ROCHA, ER; VERDI, F.; DALFOVO, D.;

OLSEN, K.; MENDES, T.; DERETTI, R.; SOARES, V.; LOBERMEYER, C.;

MOREIRA, J.; FERREIRA, J.; FRANCISCO, A. **Riscos da automedicaçâo** : tratando o problema com conhecionamento. Anais de résumos 3° seminario integrado de ensino, pesquisa e extensâo - SIEPE da Univille, 2007.

PEREIRA, Mariana Linhares; NASCIMENTO, Mariana Martins Gonzaga do. Das boticas aos cuidados farmacêuticos: perspectivas do profissionals farmacêuticos. **rev.**

Arm. Farm. 92 (4): 245-252, 2011

PORTO, CC **Medical semiology** . 5th ed. Rio de Janeiro: Guanabara Koogan, 2005.

PROPAGANDAS . **Biotômico Fontoura** . Available to:

< http://www.propagandashistoricas.com.br/2014/12/biotomico-fontoura-pele-anos-60.html >. Access em: 2 out. 2016.

. **Melhoral.** Available at: < http://www.propagandashistoricas.com.br/2014/08/melhoral-alivio-imediato-anos-50.html >. Access em: 2 out. 2016.

. **Biotômico Fontoura** . Available at: < http://www.propagandashistoricas.com.br/2014/12/biotomico-fontoura-pele-anos-60.html >. Access em: 2 out. 2016

QUALIDADE DE VIDA: um instrumento para promoçâo de saùde. **Revista Baiana de Saùde Pùblica,** v.32, n.2, p.232-240, May/August. 2008.

RAMALHO, VCS **Discourse and Ideology in Drug Propaganda** . Theses (Doutorado em Linguistica). UnB, Brasilia, 2008.

RIBEIRO, VV, et al. A sober approach to self-medication is the consumption of psychotropics in Campina Grande- Paraiba. **Infarma** . v.15, n.11-12, 2003.

ROMANO-LIEBER, N.; CUNHA, M.; RIBEIRO, E. A Farmàcia como Estabelecimento de Saùde. **Revista de Direito Sanitàrio** , v.9, n.3, 2008.

RUIZ, ME Risks of self-medication practices. **Current Drug Safety** , v.5, p. 315 323, 2010.

SA MB, BARROS, JAC, SA, MPBO Automedicaçâo em idosos na cidade de Salgueiro-PE. **rev. Arm. Epidemiol.** v.10, n., p.75-85, 2007.

SANTOS, V. back; NITRINI, SMOO Indicadores do uso de medicamentos prescriptos e de assistance ao pacientes serviços da saùde. **Saùde magazine**

Pùblica , v. 38, n. 6, p. 819-826, 2004.

SCHMID, B.; BERNAL, R.; SILVA, NN Automedicaçâo em adultos de baixa renda no municipio de Sâo Paulo. **Revista Saùde Pùblica** , São Paulo-SP, v. 44, n. 6, pg. 1039-1045, 2010.

SILVERIO, Marcelo Silva e Leite; GONÇALVES, Isabel Cristina. Qualidade das prescriçôes em municipio de Minas Gerais: uma abrondação darpuo epidemiològica, **Rev. Assoc. Med. Bras.,** Sao Paulo, v.56, n.6, p. 675-680, 2010.

SINGER, MK Redefining health: living with cancer. **Shock. Sci. Med** ., v. 37, p. 295 304, 1993.

SINITOX. Sistema Nacional de Informaçôes Tóxico-Pharmacológicas, 2009.

Available at: < http://www.fiocruz.br/sinitox_novo/cgi/cgilua.exe/sys/start.htm?tpl= home >.

Accessed: September 17, 2016.

SOARES, J. C. R. S. Quando o anùncio é bom, todo mundo compra: o projeto Monitoraçâo da propaganda de medicamentos no Brasil. **Ciên Saùde Colet** ., v.13, p. 641-49, 2008.

SOUSA, HWO; SILVA, JL; NETO, MS A Importância do Profissional Farmacêutico no Combate à Automedicaçao no Brasil. **Revista Eletrônica de Farmàcia,** Imperatriz-MA, v. 5, n. 1, p. 67-72, 2008.

SOUZA, JFR; MARINHO, CLC; GUILAM, MCR Drug consumption and internet: Critical analysis of a virtual community . **Revista da Associação Mèdica Brasileira** , v.54, n. 3, p.225-231, 2008.

SOUZA, JS; STEIN, AT Vigilância sanitària de uma cidade metropolitana do Sul do Brasil: Implantaçâo da gestâo plena e efetividade das açôes. **Ciência & Saùde Coletiva** , v. 13, p. 1-25, 2007.

TEIXEIRA, Joao Batista Picinini; SANTOS, José Vinicius dos. It is phytotherapeutic

Interações Medicamentosas . Available em: <

http://www.ufjf.br/proplamed/files/2011/05/Fitoter%C3%A1picos-e-Intera%C3%A7%C3%B5es-Medicamentosas.pdf >. Accessed: August 20, 2016.

TOMASI, E., Sant'Anna GC, OPPELT, AM, PETRINI, RM, PEREIRA, IV, SASSI, BT Condições de trabalho e automedicaçâo em professionals da rede básica de saùde da urban area de Pelotas-RS. **rev. Arm. Epidemiol.** v.10, p.66-74, 2007.

UNIVERSIDADE FEDERAL DE SAO PAULO - UNIFESP. **Xaropes e Gotas para tosse.** Available at: < http://www2.unifesp.br/dpsicobio/cebrid/folhetos/xaropes _.htm. Acesso em: 21 sets. of 2016.

VARELLA, Drauzio. **Anti-inflammatory.** Available em:< https://drauziovarella. com.br/letras/a/anti-inflamatorio/ >. Acesso em: 21 sets. of 2016.

VIEIRA, AA Laboratòrio de Anàlises Toxicológicas de Emergência em Hospitais. **rev. Inter** . v.5, n.3, p.60-75, 2012.

VITOR, RS; LOPES, CP; MENEZES, HS; KERKHOFF, CE Pattern of drug consumption sem prescriçao medica in the city of Porto Alegre, RS. **Revista & Saùde Coletiva** , Porto Alegre-RS, v. 13, Sup I, p. 737-743, 2008.

OLIVE, MPA; BARROS, MBA; CÉSAR, CLG; CARANDINA, L.;

GOLDBAUM, M. Hipertensao arterialem idosos: prevalence, associated factors and control practices in the municipality of Campinas, Sao Paulo, Brazil. **Caderno de Saùde Pùblica** , v. 22, p. 285-294, 2006.

WERTHEIMER, AI; SERRADELL, J. A discussion paper on self-care and its implications for pharmacists. **Pharm Wordl Sci** . v.30, n.4, p.309-315, 2008.

WOLF, MM; LICHTENSTEIN, DR; SINGH, G. Gastrointestinal toxicity of monosteroidal/ anti-inflammatory drugs. **N. Engl. J. Med.** , v. 34, n. 18, p. 88-99, 1999.

7   APPENDICES

ANALYSIS OF THE

SELF-MEDICATION PROFILE OF RESIDENTS OF THE

CANGALHEIRO, SALOBRO AND SERIEMA NEIGHBORHOODS IN THE CITY OF CAXIAS-MA

Supervisor: Prof. Me. Joao Luiz Macedo de Sousa Cardoso Researcher: Ana Florise Morais Oliveira

This information is being provided for the voluntary authorization of your participation in this research, which aims to analyze the risks of self-medication among residents of the Cangalheiro, Salobro and Seriema neighborhoods in the city of Caxias, MA. The research will be qualitative and quantitative, as it seeks to understand and demonstrate the data analyzed through statistics.

To collect the information we will use a written questionnaire and also questions asked orally, through a conversation with the interviewee about the topic.

The research offers minimal risk to the participant, and may cause discomfort and/or embarrassment when addressing self-medication.

By collaborating with the research, the subject will benefit directly and indirectly, because by talking about the practice of self-medication, they will provide guidance on the research, contributing to greater knowledge on both sides.

Researcher Ana Florise Morais Oliveira and advisor Joâo Luiz Macedo de Sousa Cardoso can be reached at: (99) 982465800, aflorise@gmail.com and (86) 99225937, jlmacedosousa@gmail.com, respectively.

Therefore, by allowing the minor to take part in the research, you will have the freedom to withdraw from it at any time, without any prejudice. You are free to withdraw your consent as soon as you decide to do so.

All information will be kept confidential and the researcher will be completely trustworthy. There will be no personal expenses for the participant, and they will not receive any financial compensation related to their participation, i.e. they will participate voluntarily.

INFORMED CONSENT

Me,   ,

RG I agree to participate in a study of the following nature

The aim of this study is to analyze data through a quantitative and qualitative survey of 900 people who use self-medication in the Cangalheiro, Salobro and Seriema neighborhoods in the city of Caxias-MA, using an individual field research questionnaire developed and applied by Ana Florise Morais Oliveira, an undergraduate student in Biomedicine at the Mauricio de Nassau College, under the guidance of Professor Joâo Luiz Macedo de Sousa Cardoso.

I also state that I am aware that the information will be treated anonymously and confidentially. Furthermore, I am aware that my participation in the research will not involve any risk or harm of any kind, and I have also been informed that, as it is voluntary, it is not obligatory, and I have no financial interest.

I also authorize the use of the data from the interview I gave to obtain the results of the study carried out and affirm that I am faithful and truthful in my answers.

I declare that I have understood the objectives and conditions of my participation in the research and agree to take part.

____

(Signature of research participant)

ID/CPF

____

Name and Signature of Responsible Researcher

ID/CPF

Caxias (MA), ________2016 ____________________ .

APPENDIX B - QUESTIONNAIRE APPLIED

INTEGRATED CENTER FOR HIGHER EDUCATION IN PIAUI

MAURICIO DE NASSAU COLLEGE/ALLIANCE UNIT

bachelor's degree in biomedicine

ANALYSIS OF THE

SELF-MEDICATION PROFILE OF RESIDENTS OF THE

CANGALHEIRO, SALOBRO AND SERIEMA NEIGHBORHOODS IN THE CITY OF CAXIAS-MA

Researcher: Ana Florise Morais oliveira

Advisor: Prof. Me. Joao Luiz Macedo de Sousa Cardoso

OBJECTIVE: To analyze data through a quantitative survey of people who self-medicate in the Cangalheiro, Salobro and Seriema neighborhoods.

SOCIO-ECONOMIC QUESTIONNAIRE

Name: ____

Address: ____

Profession: ____

gender: Male (   ) Female (   )

1) Age (years) between:

a) 18-25 years ( )

b) 26-30 years ( )

c) 31-40 years ( )

d) >40 years ( )

2) its color is:

a) White (   )

b) Black (   )

c) Brown (   )

d) Indigenous ( )

e) Other (   )

3) Level of education:

a) Illiterate (   )

b) semi-literate (   )

c) Complete elementary school (   )

d) Elementary school incomplete (   )

e) High School (   )

f) university degree completed ()

g) Postgraduate (   )

4) What is your current family income?

a) Up to 1   Minimum wage   (   )

b) 2 a4salaries   ( )

c) 3 a5salaries   ( )

d) 5 a7salaries   ( )

e) ≥8 salaries (   )

5) Individual income:

a) Up to 1 minimum wage (   )

b) 2 to 4 salaries ( )

c) 5 to 7 salaries ( )

d) ≥8 salaries (   )

e) I'm not employed ( )

6) You benefit from a government program:

a) Bolsa familia (   )

b) Scholarship (   )

c) PROUNI (   )

d) FIES (   )

e) My house My life (   )

f) Light for All Program (   )

7) What kind of media do you have the most access to?

a) TV (   )

b) Radio and telephone ( )

c) Internet ( )

d) Newspapers (   )

e) Magazines (   )

f) Social media (   )

8) You use tobacco or someone else in your household:

a) Smoker (   )

b) Non-smoker   ( )

c) Ex-smoker (   )

d) Another person in the   house is a smoker (   )

9) Do you practice any form of physical exercise?

a) Yes, less than 3 days a week (   )

b) Yes, between 3 and 6 days a week ( )

c) Daily (   )

d) I don't do any kind of physical activity (   )

10) Have you experienced any symptoms of discomfort in your body with the use of any medication in the last three months?

a) Yes ( )

b) No (   )

c) Which symptom? ____

11) This medication(s) was (were) prescribed by a doctor: a) Yes (   )

b) No (   )

12) What types of medication do you use most in your day-to-day life?

a) Anti-inflammatory ( )

b) Analgesic / Antipyretic (   )

c) Vitamins (   )

d) Antigripal (   )

e) Antiallergic ( )

f) Antiparasitic (   )

g) Antidepressant (   )

h) Cough sedative (   )

i) Antihypertensive ( )

j)   Food supplement (   )

k)   Contraceptive (   )

l)   Other: ____

13) How often do you use this medicine?

14) Do you self-medicate?

a) Yes ( )

b) No (   )

15) Why did you decide to self-medicate?

a)   Influence of the media (   )

b)   You've taken it other times ( )

c)   Difficulty accessing a doctor (   )

d)   No time (   )

e)   Was guided by other people (   )

f)   Other (   )

16) Do you use self-medication to solve any health problems? a) Yes ( )

b) No (   )

17) What were the reasons for taking the medication?

a)   Headache (   )

b)   Hepatic colic (   )

c)   Abdominal colic (   )

d)   Poor digestion ()

e)   Cough ( )

f)   Infection ( )

g)   Allergy ( )

h)   Muscle pain (   )

i)   Appetite stimulant (   )

j)   Appetite suppressant ( )

k) Anxious (   )

l) Chronic illness   ()

m) Other illness (   )

18) These medicines come from: a) Natural products (   )

b) Industrialized (   )

c) Homeopathics (   )

d) Herbal medicines (   )

19) Can you say how many different types of medication you have used in the last six months?

a) 1X( )

b) 2X( )

c) 3X( )

d) 4X( )

e) 5 or more (   )

20) Has this type of medication been used for the same illness?

a) Yes (   )

b) No (   )

21) How many times have you been to the doctor in the last 12 months?

a) 1X ( )

b) 2X (   )

c) 3X ( )

d) 4X ( )

e) 5 or more (   )

22) Have you had any clinical and/or laboratory tests?

a) Yes ( )

b) No (   )

23) Do you know the risks of the side effects of the medication you use?

a) Yes ( )

b) No ( )

24) Have you ever bought any kind of medicine from a drugstore?

a) Yes ( )

b) No ( )

25) When you bought the medicine at the drugstore, did the clerk ask you for a prescription?

a) Yes ( )

b) No ( )

26) Do you believe that the pharmaceutical sales system in Brazil works in accordance with the law?

a) Yes ( )

b) No ( )

8 ANNEXES

ANNEX A - MAPS OF THE NEIGHBORHOODS SURVEYED

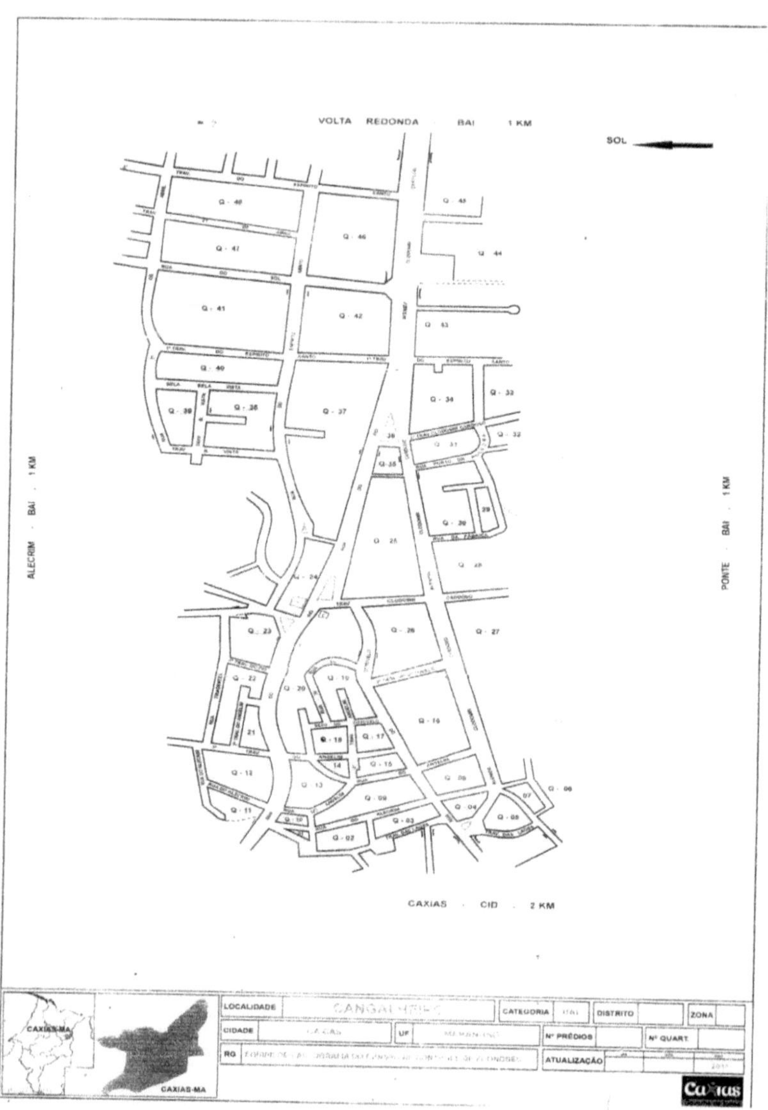

Seriema neighborhood

Source: Caxias City Hall, 2016.

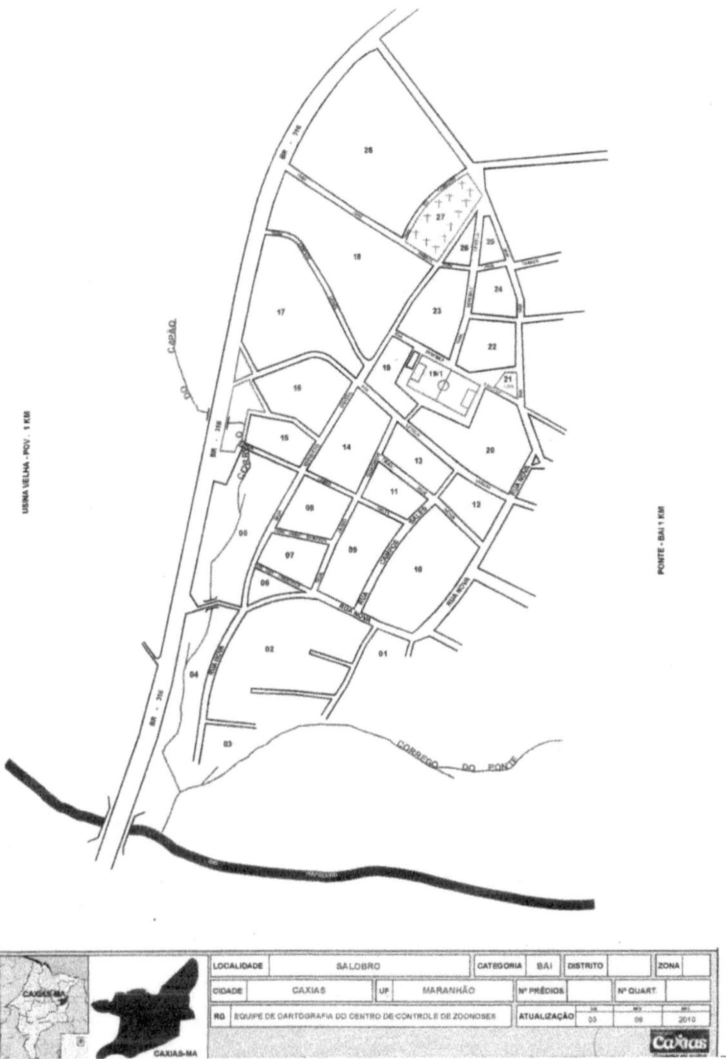

Seriema neighborhood

Source: Caxias City Hall, 2016.

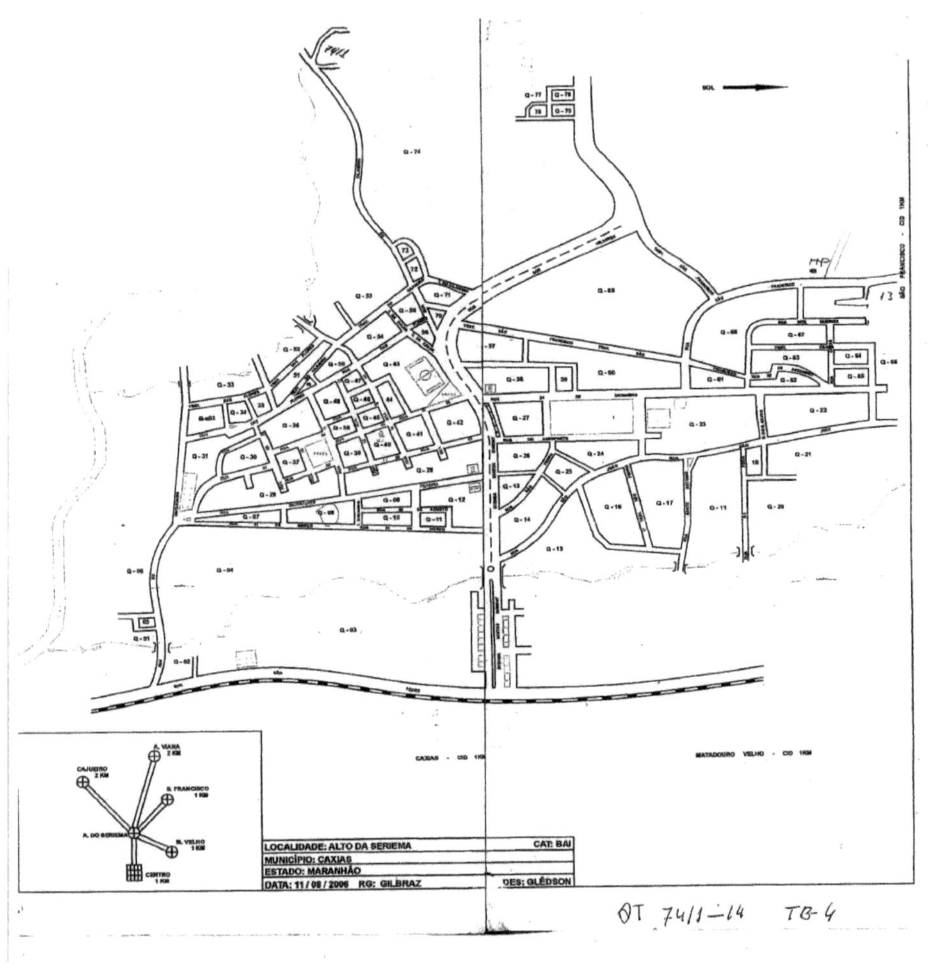

Seriema neighborhood

Source: Caxias City Hall, 2016.

I want morebooks!

Buy your books fast and straightforward online - at one of world's fastest growing online book stores! Environmentally sound due to Print-on-Demand technologies.

Buy your books online at
www.morebooks.shop

Kaufen Sie Ihre Bücher schnell und unkompliziert online – auf einer der am schnellsten wachsenden Buchhandelsplattformen weltweit! Dank Print-On-Demand umwelt- und ressourcenschonend produzi ert.

Bücher schneller online kaufen
www.morebooks.shop

info@omniscriptum.com
www.omniscriptum.com

Printed by Books on Demand GmbH, Norderstedt / Germany